HOMESTEADING FROM SCRATCH

HOMESTEADING FROM SCRATCH

Building Your Self-Sufficient Homestead, Start to Finish

STEVEN JONES

Skyhorse Publishing

Skyhorse Publishing books may be purchased in bulk at special discounts for sales promotion, corporate gifts, fund-raising, or educational purposes. Special editions can also be created to specifications. For details, contact the Special Sales Department, Skyhorse Publishing, 307 West 36th Street, 11th Floor, New York, NY 10018 or info@skyhorsepublishing.com.

Visit our website at www.skyhorsepublishing.com.

10 9 8 7 6

Library of Congress Cataloging-in-Publication Data.
Names: Jones, Steven, 1976- author.
Title: Homesteading from scratch : building your self-sufficient homestead, start to finish / Steven Jones.
Description: New York : Skyhorse Publishing, [2017] | Includes bibliographical references and index.
Identifiers: LCCN 2016043115 | ISBN 9781510712904 (pbk. : alk. paper)
Subjects: LCSH: Agriculture. | Home economics, Rural. | Country life. | Urban homesteading.
Classification: LCC S501.2 .J66 2017 | DDC 635–dc23
LC record available at https://lccn.loc.gov/2016043115

Cover design by Jane Sheppard
Cover photo credit: bottom row, Melissa Jones; top photo, iStock

Print ISBN: 978-1-5107-1290-4
Ebook ISBN: 978-1-5107-1294-2

Printed in China

Dedication

This book is dedicated to my wife, Melissa,
and my children, Oliver and Hannah.
They are the only reason I do anything.
This book is also dedicated to the memory of
my grandfather, Clyde "Beck" Jones,
who I wish could have lived long enough to see it.

Contents

PREFACE

Homesteading From Scratch was designed to explore homesteading with readers who have no prior homesteading experience or knowledge.

This book is for people who want to do things differently. The type of people who want to eat real food, grow herbs, make cheese, raise baby animals, hunt mushrooms, pick blackberries, "unschool" their children, can jelly, ferment kraut, farm organically, connect to nature, live intentionally, and more.

Touching on farming, animal husbandry, home production, food preparation, and even homeschooling, *Homesteading From Scratch* allows readers to discover homesteading as a movement and as a lifestyle.

Inspired by *From Scratch* magazine, an online publication devoted to homesteading and intentional living, this book serves as a reference and also as a cheerleader for people who want a bit more control and responsibility over where their food comes from, the things they consume, and how they live their lives.

People like Katherine and Bobby Benoit, who've been homesteading for twelve years. They produce 80 percent of their own food and homeschool their seven children. Or Randall and Elizabeth Wescott Hewitt, who raise heritage pigs and cattle on their family property. There's Tiffany Toler of the Cedar Roost who has been working to achieve her dream of creating an all-natural, organic hatchery. And then there are people like Tiffany Ketchum, who's trying to turn her green thumb and vegan lifestyle into a self-sufficient dream in her backyard while working on her degree and juggling life as a full-time mother of two gorgeous children.

Some of us, like Katherine and Bobby, have been living a self-sufficient life for a long time. Others, like Tiffany, have just begun exploring a homesteading lifestyle and developing a base skill set.

That's what's really great about this movement. There are thousands of people who want to live a life "close to the ground," and would love to work with you to develop even more of a community by sharing skills, knowledge, and ideas. I'm blessed to be part of this group of people who are devoted to learning and passionate about their surroundings.

When we became homesteaders, my family and I made friends all over the world who are devoted to healing themselves, their families, and their communities through something as simple as learning to live a more intentional life through raising food, animals, and babies.

Hopefully, this book can serve as an aid to those people who just want to do things differently.

Thank you and enjoy!

INTRODUCTION

My grandparents were born in Alabama in the 1920s, and both were children of sharecroppers. My grandfather, Clyde "Beck" Jones, always told us stories about growing up and how he started plowing fields with a mule at the age of thirteen. My grandmother, Inez Jones, beat him by a few years. She cannot remember how old she was, but she recalls having to stand on tiptoe to reach the handles of the plow.

They spent decades growing cotton, peanuts, and tobacco. They grew up in homes that were heated by wood, raised most of their food in home gardens and butchered their own animals whenever they needed meat.

They continued these habits well after they left farming to get jobs at the cotton mills. My grandmother kept a "kitchen garden" that spanned more than an acre and included fruit trees and blueberry bushes. Honestly, I believe the woman could spit on the red Alabama clay, throw down a seed, and produce enough food to feed a family of twelve, like some sort of Biblical miracle.

She froze, canned, and dried so much food, that even now, in her late eighties, she'll probably have enough jars put up when she leaves this world (may it be a blessedly distant day) to feed a small army. This is despite the fact that she quit canning years ago.

In contrast, my parents worked in factories and in construction. They kept a small garden only sporadically. My mother occasionally made jelly and canned a few odds and ends; however, the rigors of raising four rambunctious children made the idea of "going back to the land" seem like something for people with a lot more spare time.

I vaguely remember hoeing in my grandmother's garden for the last time when I was fourteen. A surly teenager, I was unhappy about the whole experience.

I should have spent every spare second in my saintly grandmother's garden trying to absorb as much as I could while she was still healthy enough to show me. But I didn't.

Like many people, I grew up with only the vaguest of ideas about where my food came from. I grew up in rural Alabama, so I had a better idea than some, but despite my parents' best efforts (my father once slaughtered and butchered a hog in my backyard) I knew (and know, if I'm being honest) only the barest minimum about growing food and living off the land.

So, when I was in my late thirties, I started thinking about farming. Truthfully, it started off as a joke. I was leaving a job in television and wanted to get as far away as possible from working in media. I knew only a few things: I was tired of living in a suburban house I didn't want; I was tired of being in a job that involved nothing but a computer; and I was tired of working just to afford things I did not get to enjoy most of the time.

I like animals, and I like working with my hands, so the idea of doing something that involved both of those things appealed to me.

They appealed even more to my wife. When I told her (again, only as a joke) that I'd like to start a goat farm after leaving television, she ran with the idea, much to my surprise.

We started researching and thinking about how we lived our lives.

We determined that we wanted to know more about our food and where it came from. We determined we wanted to live in a more self-sufficient fashion. We made the decision to start a farm, not a goat farm, but something a bit more holistic. That's when we discovered the homesteading movement.

Visions of *Little House on the Prairie* ran through our heads. We fantasized about living in the wilderness, clear-skinned and attractive, while our above-average children ran through flower fields in homespun clothing.

That was not the reality, and we are happier for it.

There is a lot more work (Pa never seemed too sweaty on the television show) and a lot more know-how involved.

There's a lot of information out there, and it's all over the place.

We found out there's a big argument over what "organic" is, and whether or not it's important.

We learned that there's dozens of different ways to "homestead" (for example, did you know that you can homestead in an apartment?) and at least as many different ideas on what it meant.

So, we did what any creative couple with about forty years of combined media experience would do: we started an online magazine, *From Scratch*.

Our goal, even now, is to collect as much information and ideas about homesteading as possible, do the best we can to make the movement as accessible and attractive to as many people as possible, and learn as much as we can while we do.

When we started our journey, we were living in a suburban house that was big enough for three families. Now we live in a house just big enough for us, lease a five-acre micro farm, and ferment everything we can think of.

We've milked goats, slaughtered chickens, grown way too much squash, baked bread, taken up crochet, incubated chicken eggs, and homeschool both of our beautiful children.

In the process, I've dealt with crop failures and animal deaths and gotten animal poop in my mouth more times than anyone ever should.

We've learned a lot, but more than anything we've learned how little we do know.

Like many people in this country, we're a generation removed from skills and abilities that children used to learn from birth, like my grandparents did.

While we have no desire to go back to sharecropping commodity crops—there's a reason my grandparents took factory jobs the first chance they got—we do feel like we as a culture may have lost something in moving toward massive industrial farming and away from smaller, decentralized food production.

This book is a continuation of those beliefs and our desire to make as much information as possible available to as many people as possible.

While there's a ton of information on homesteading, farming, food, homeschooling, etc., there's still a gap on how to achieve the "homesteading dream," whatever that may be. The assumption of most "experts" is that people either know nothing about what they're doing, or they already know everything they need to know.

Bearing that in mind, the point of this book is to show you how to get from desire to reality, armed with nothing but the information gathered from these pages.

With this book, it's my hope that you decide to take up homesteading, read everything you can, and start doing.

I've mapped a path that you can follow to achieve your *Little House on the Prairie* dream, no matter what it looks like.

Good luck, have fun, and watch out for the poop!

1 | Section One: Getting Started

CHAPTER 1: WHAT IS HOMESTEADING?

A working definition

Homesteading means many different things to many people. Historically speaking, the term came into use regarding the various Homestead Acts in 1862. Essentially, the term, as used then, described small, self-sufficient farmers and communities. Before that, the terms "smallholder" and "yeoman farmers" described similar concepts.

Now, depending on who you are and where you're located, it seems "homesteading" can mean anything a person wants it to.

There's nothing wrong with that. However, for the sake of this book, we need a definition—something to work with.

For us, homesteading will refer to the lifestyle devoted to self-sufficiency with an emphasis on home production and responsible consumption. In plain English, this means people who engage in homesteading are determined to make their own way— to do whatever they can while depending as little as possible on mass-produced items, be it food, clothes, or household goods.

These are people who raise chickens in their backyard, use a sewing machine, bake their own bread, and plant a garden. Homesteaders are canners and DIYers, fence-building, modern-day pioneers who have removed themselves from the rat race. They wear boots and jeans and love beautiful, handmade things. They enjoy good music, good food, and good books.

Why should you homestead?

If you like those things, you might want to consider homesteading. But be aware, this lifestyle is difficult.

It's a lifestyle devoted to doing things the hard way. In fact, it's almost a badge of honor among homesteaders to see who can find the hardest way to do something.

Instead of buying milk, a homesteader would rather milk a goat. But, since it's cruel to have less than two goats, you decide to get two goats. Those two goats must be bred regularly, and you'll work really hard to find a buck (a male goat). Then you'll decide to buy a buck. Then you'll breed your two does (a doe is a female goat). Those does will produce two kids each and you'll find yourself unable to part with the babies, as baby goats are possibly the most adorable creatures in the world. So, now you have a herd of goats (seven total), you spend hours a day milking, feeding, providing medical care and more to your goats. It costs you about $300 a month just to buy food for the animals. Not to mention a good milking goat can run you about $300. But,

If you start out with two goats, you may end up with a herd, since baby goats are possibly the most adorable creatures in the world.

now you don't have to spend four dollars on a gallon on milk, as you get about half a gallon of milk a day from the two mature milking goats!

Essentially, if you're a person who is more interested in the journey than the destination, then homesteading is probably for you.

What are you homesteading for: money, self-sufficiency, passion?

While all homesteaders aspire to have some degree of self-sufficiency, people choose this lifestyle for different reasons. Some are preppers, determined to be able to take care of themselves and their families. Others are just fed up with a centralized food system and want to do their part to make locally produced food and all that represents more widely available. Still others surprisingly go into homesteading as a small business (perhaps even more surprisingly, many of them are successful at it). No matter what your reasons, take a moment to reflect on why you'd like to start homesteading. Writer it down. Make a vision board, a brain map, a to-do list, whatever it takes. This time of reflection will go a long way toward getting you in the right mindset. Without at least some idea of why you've decided to take on a life of self-imposed difficulty, then you might wind up in a situation that's too chaotic—and expensive—which will inevitably lead to quitting. And while there's nothing wrong with quitting something that doesn't work, we need all the homesteaders we can get.

Different ideas about homesteading

So, there's different ways to deal with homesteading—different levels of commitment, if you will. (Note: One level is no better or worse than the other. Our unofficial motto at *From Scratch* is "We're all at different places on our own paths." Some people involved in the homestead lifestyle believe in litmus tests. We don't.)

You can start an urban homestead, where a home in a city provides you with all the space you need to grow food and raise chickens, and puts you close to any market you'd like to sell in.

Also, with newer farming models, like SPIN (**S**mall **P**lot **IN**tensive) Farming, Farm-A-Yard, and biointensive farming methods, a lot of food can be grown in very small spaces with very little equipment.

Urban homesteading has one big advantage, as it helps alleviate the social isolation that many small farmers, to their detriment, suffer from. However, for many, urban homesteading doesn't allow for the space and self-sufficiency some require for their personal homesteading journey.

Which brings us to the "traditional" homestead, or what people think of when they hear the word "homestead": the farmhouse out in the country.

Whether purchased or built, this model offers the *Little House on the Prairie* dream many homesteaders have. It has a lot of advantages. Greater space means more animals, plants, and projects, which can increase profits and self-sufficiency. It's great to raise children out in the country, away from the hustle and bustle of the city. If you homeschool (more on that later), then there's lots of learning opportunities to be found when you live a bit closer to the wildness of nature.

But, it can be isolating, it's often more difficult to get materials and resources to and from markets, and you may find yourself competing with farmers and operations that view a thousand acres as a "small" farm. However, rural people can usually find people with certain skills that you might have difficulty finding in urban areas. For example, if you're into fiber arts, a senior center in a rural county will undoubtedly be overrun with seamstresses who can put you on the right track. Fifth-generation farmers in the boonies can tell you all about which varieties of okra to grow and the best place to find seed and feed.

My favorite version of homesteading, so far, however, is apartment homesteading. A term I've only heard used by Samantha McClellan, of the *Sweet Potatoes and Social Change* blog, I'm hoping apartment homesteading catches on. Samantha lives in a pretty small apartment but manages to live a homesteading lifestyle despite her limitations. She composts, raises worms, gardens in containers, air dries her clothes, and so much more. It's a great method to get started in homesteading and it's a great way to introduce yourself to homesteading concepts and principles before you run off and grab a thirty-year mortgage.

Different homesteading goals

Now that you know some of the different ways to homestead, you should think about your homesteading goals. Knowing what you want to achieve with homesteading will make a difference in the type of homesteading you'd like to do. For example, if you want to start a dairy, then having a cow in your apartment homestead probably isn't the way to go.

As I mentioned before, every homesteader has different reasons, and as you determine what your reasons are, you'll need to set goals before you take the leap.

Some homesteaders earn thousands every year with their homesteading business. Some homesteaders need to have a fully functioning farm to feel like they've "made it." Others get a coop and chickens in the backyard and they're finished.

Again, all of these are valid goals, but you'll need to figure out what *your* goals are before making any determination on how and where you'll be homesteading.

Determining your goals

The below steps will help you establish your reasons for homesteading, what type of homestead you want, and what your homesteading goals are.

Step 1: Brainstorm

Grab a pencil and paper and start jotting things down. As you write, try to think about why you'd like to change your life. Maybe food is important to you and you'd like to know more about where it comes from. Maybe you're an animal lover and you'd like to spend more time with feathered and furry friends. No matter what your reasons, have fun with it. Get a little starry eyed and dream big.

Step 2: Envision your homestead

I'm a big fan of vision boards, and this is the perfect opportunity to make one. For those who don't know, a vision board is a visual representation of what you're trying to achieve. Most of the time, vision boards wind up looking like collages (Remember those from school?) with pictures cut from magazines and pasted onto poster board (pro tip: glitter glue is your friend) but they can be anything: Pinterest boards, a list on a piece of paper, a carefully curated slideshow on your website, anything. The point is, you can take this and determine what your homestead will look like, allowing you great insight into how you'll create that space. You need something you can look at, visualize, and refer to often. Again, have fun and dream big!

Step 3: Determine your goals

This is where you really start to formulate your plan. Now that you know why you're homesteading and how you want to homestead, you can set achievable goals. These can be as simple or as complex as you want, but you'll want to make sure you have a deadline. So, put together three lists: one year, three years and five years.

Each list will contain goals to achieve to get you where you want to be on your homesteading journey. Something like:

Year one: By the end of the year I'll have a chicken coop, consume 20 percent of my food from local sources, and have a garden growing.

Year three: By the end of the year I'll produce 20 percent of the food I eat from my homestead, have a flock of twenty birds, and be selling eggs at the farmers' market for extra income.

To turn your dream of homesteading into a reality, you need to set some goals.

Year five: By the end of the year I'll be living in a yurt, have a few acres producing veggies for the farmers' market, and selling meat from my chicken flock.

Ultimately, this process should be fun. You're at the beginning of a big journey, but with a little forethought, you can give yourself the best chance for success. And this game is rigged for success, as only you can determine what it looks like in your unique situation. Moving forward, remember to move at your own pace and comfort level and to start small. This is important, as homesteading can be a big financial commitment and major life change for you and your family. You don't want to wind up in a situation where you're uncomfortable and saddled with debt only to discover it's not for you.

Homesteader profile: Samantha McClellan

Samantha McClellan, of the *Sweet Potatoes and Social Change* blog, is a homesteader in North Carolina. She's been an apartment homesteader since 2013. Her journey actually began with a health issue. In 2011, she was diagnosed with ulcerative colitis, which led her to explore the Paleo Diet. This led to her belief that food could be medicine—powerful medicine in her case—leading her to look into ways to advocate for a better outcome for herself and others. In addition to homesteading and blogging, Samantha's also a doula and active in the natural childbirth movement.

"I didn't want to use homesteading as a means of hiding or backtracking to the 'good old days.' Rather, I wanted to use it as a means of embracing the present and advocating for a better future, a future that fused the wonder of technological advancement and modern medicine with the tried and true wisdom of nature and past generations," she explains.

Find out more about Samantha at: www.sweetpotatoesandsocialchange.com.

Is homesteading for you?

To help determine your comfort level, you may want to try out some things first. Schedule a retreat at a commune, take a workshop on meat processing, do a farm internship, or just take a few classes at your local cooperative extension office. These are easy to find and lots of fun, so even if you decide that homesteading isn't for you, or that your comfort level is lower than you thought, you'll still meet some great people and learn a lot.

CHAPTER 2: WHAT DO YOU BRING TO THE TABLE?

Assessing abilities

When you first decide that you want to homestead, you need to sit down and figure out what you have in your favor.

You have to put together a complete rundown of your resources (land, equipment, cash, labor, etc.) and your abilities (the things you know how to do, your physical ability to operate equipment, even your general physical health).

This list will help guide you in a direction that allows you to make the most of what you have and what you can do. Without it, you can easily wind up in over your head. You don't want to run out and spend five figures on a tractor if you don't know how to use it, or you don't have land to operate it on.

Start with an assessment of your abilities, and, again, this is a situation where there are no wrong answers. This can be broken down into a few different categories: your knowledge, your physical capabilities, and any miscellaneous skills.

For example, when it comes to your physical abilities, you'll want to know how long you can spend outdoors. Do you have hiking or camping experience? Are you a runner? Are you a weekend warrior? You don't have to be a high-functioning athlete with tons of experience as a rock climber, but you do need to be realistic about what you can and cannot do. Just like not running out and buying a tractor is a bad idea when the lot you live on is one-tenth of an acre, running out and deciding to work a two hundred-acre piece of land covered in trees is a bad idea if you haven't built up the strength to handle a chain saw.

Don't worry though, if your dream is a two hundred-acre wild and woolly lot, then you can figure out a way to build up your strength and use tools and outside labor to achieve what you want. For now, however, just make a list of your physical capabilities so that you can plan around it. It's possible to homestead even if you cannot bend at the knees, but if you ignore the fact that you cannot bend at the knees, then you will get frustrated early and often. Frustration causes homesteaders to quit, and we want you to have every chance of success.

Once you're done listing your physical abilities, you'll want to make a list of what you know. Do you know how to grow vegetables? Are you a knitter? Do you know what to feed chickens? Do you know about the basics of soil science?

Anything you know how to do, or know enough to figure out, you'll want to put on this list. This gives you a great starting point for your personal homesteading dream.

With this list, you can extrapolate abilities. If you can knit, you can expand your knowledge of the fiber arts into spinning, weaving, crochet, and even

If you know the basics of chicken keeping, you can set up a backyard coop and start raising birds for meat and eggs.

keeping sheep and angora rabbits. If you know the basics of chicken keeping, you can set up a backyard coop and start raising birds for meat and eggs. If you know how to grow vegetables, then you can learn how to grow enough food to sustain yourself and your family.

This list will also give you a starting point on what you need to learn and how to expand on that knowledge. If you can keep chickens, you can learn how to keep and raise ducks. If you know the basics of soil science, you can learn about ways to increase beneficial microbes in your soil. And don't think some tidbit of information or random hobby you have won't come in handy. If you brew your own beer for fun, then you'll have a leg up when it comes to fermenting food and making kombucha.

Finally, make a list of any miscellaneous skills you have. Forklift operator, shift manager at a fast food restaurant, whatever. These are those weird little things we all pick up in life that are a big, big help later; for example, I can fieldstrip a toilet blindfolded.

Again, just like your knowledge base and your hobbies, you'll want a reckoning of what you're able to do and how well you can do it. Before I forget, you'll want to make a list for everyone in your household. Not only will it give you guidance on what kind of homestead you want, but it will also help you divide the labor. If your daughter's a whiz at running a cash register because of her summer job at a local grocery store, then put her in charge of the till at the farmers' market when you start selling crafts and veggies.

With planning and preparation, you can grow plenty of food and keep animals on almost any size property.

Again, have fun with this exercise. Don't be afraid to put something on the list that doesn't make sense. Then you and your family can get together and make plans, laugh, and dream together. It'll be such a fun time.

Assessing resources: do you need land?

Now that you know what you and your family members can do, and what each of you brings to the table, now's the time to determine whether or not you need land. The answer, believe it or not, is probably no.

Having a homestead, in this day and age, is just as much about a lifestyle as it is about a location.

As mentioned in Chapter 1, you can homestead in an apartment if you want to. And with the proper preparation and plan, you can grow plenty of food and keep animals on any size property.

Bees can be kept on rooftops. Chickens can be kept in window enclosures. Check out Chris McLaughlin's book, *Vertical Vegetable Gardening,* to get all sorts of great ideas on how to maximize space.

There are homesteaders producing thousands of pounds of food on small urban plots. There's no limit to what can be done with the proper preparation and effort. There are even some municipalities that allow residents to keep milk goats because the animals are legally considered pets.

So, if you're interested in homesteading, the answer to whether you need land is probably no. But whether you want land or not is a different consideration all together.

Be aware, if you decide to run out and find a nineteenth-century farmhouse on forty acres in Vermont, that's going to be a considerable cost. There's nothing wrong with taking on a mortgage for your dream property, but don't expect your fledgling homestead business to cover your house payment for the next thirty years.

Note, you can also upgrade at any point in time. It's a lot easier to buy more property than it is to sell it if you find yourself in a situation where you've overextended your credit and over-leveraged your income. Being saddled with an upside down mortgage, despite what you may think of your dream property, is not homesteading, as it violates the principles of self-sufficiency. Even if you decide that taking on more debt is something that you're able to do, don't jump into any long-term debt agreement without serious thought and long-term planning.

Another thing to be aware of is this: land is cheaper to lease, especially farmland.

In North Carolina, the average cost of leased farmland is $250 per year. It's not much more expensive than that in most parts of North America. And, if you're determined and looking in the right place, you can lease space in community gardens, go into a share arrangement with animals—I know several families that entered into an agreement to share a milk cow—and so much more. Many areas all over the country have farm incubator programs, incubator kitchens to start foodstuff businesses, and more.

While a lot of those different plans hinge around the infrastructure in your region/city/county, there's still a lot to be done in areas that aren't as populated and "with it" when it comes to the homesteading/local food movement. There may be an elderly couple in your neighborhood that cannot take care of their lawn anymore, so you could make an agreement with them that you get to plant vegetables in their backyard in exchange for keeping the front yard grass trimmed.

Most people, if approached with kindness and honesty, are usually willing to help out their neighbors.

A recent movement, the Farm-A-Yard organization, has a program designed specifically for small-scale farmers to raise commercial crops instead of grassy lawns. You can find out more at farm-a-yard.com.

Notes on zoning and HOA rules

As you make the determination on buying, leasing, or whatever, be aware: zoning regulations and homeowners' association's rules may not be the most hospitable where you live. Before you start tearing up your fescue lawn to replace it with rows of heirloom bean plants, make sure you're not violating some city regulation or HOA rule.

If you do wind up on the bad side of these regulations, you could be in for an abundance of legal fees. If you happen to challenge these regulations in court, you might win, but you'll spend more on your attorney that you would ever gain.

The best thing to do is see what is allowed in your neighborhood and work within those parameters, if possible. If not, then work to change those regulations. With an increased awareness of food security and the farm-to-table movement, many city council members are willing to hear your requests for changes in zoning regulations and laws, especially if you work with your neighbors to craft rule changes that benefit everyone while remaining respectful (e.g., hens only for backyard chicken coops). HOAs are notoriously difficult to change, but not impossible. You may have to run for a spot on your HOA board in order to bring about positive change. Again, remember that no matter what, requests made with kindness and respect for your neighbors will have a lot better chance to become the new "law of the land." You might also consider working with your city and neighborhood leaders to establish a community garden space. There's more than one way to bring about change. And, of course, you can always rent space in a different area to raise your chickens, veggies, bees, etc.

You will never be an expert, so you may as well get started

Let's assume, for the sake of argument, that once you're done with your list of abilities, knowledge, and resources, you look it over and see it woefully empty.

While I'd encourage you to try again, because most people have something that they can build on, you have to understand something about homesteading: There's no such thing as an expert.

A Canadian documentary, *To Make a Farm*, featured several first-generation farmers. One of the farmers was a woman who raised heritage pigs, chickens, and sheep. She had little to no prior experience in farming and ranching before embarking on her adventure.

In the film, however, she mentioned that it didn't bother her. To paraphrase, she pointed out that there's no such thing as a farming expert. If a person's lucky, they'll bring in, say, thirty crops in their lifetime. She pointed out there's nothing a person can do just thirty times and be considered an expert.

Homesteading, which can be a form of farming, is exactly like that. There's no way you'll be able to learn enough about all the skills a homesteader practices to be considered an expert on any of them.

So, embrace your ignorance and get started.

The best way to learn homesteading is by doing it. Read tons of books, ask lots of questions, get your hands dirty, and make mistakes. If you wait until you know all there is you want to know, you'll never get started.

Helpful Skills to Learn

While you're assessing your skills, you might want to put together a list of things to learn. Here are some things I started to learn when I began homesteading.

Engine repair—Most of the tools we use are hand tools or manual drive (even our lawn mower), but occasionally we'll rent or borrow a tiller. Invariably, something happens that requires me to break out a wrench or screwdriver and figure the thing out. Knowing something about engine repair is invaluable. If you decide you need a tractor for your homestead, then maintaining and repairing it yourself is almost inevitable. Pick up some books or check with your local community college. Many offer continuing education courses on engine repair. And while I'm pretty sure your local community college doesn't offer Tractor Repair 101, even a small engine class would be helpful.

Animal health care—You'll probably specialize in whatever animals you decide to keep, but you'll still need to learn some basic vet care, like how to administer injections and basic first aid. Vet bills are expensive, and there's not always a guarantee that your vet will know the ins and outs of your animals (your llama might be the only llama she's ever seen). So start talking and start reading. Breeders and herdsmen are a font of knowledge.

Plumbing—Any and all construction skills are good to know, but especially plumbing. Crops and animals depend on getting water to where it needs to be. Perhaps more importantly, you depend on getting water away from where it shouldn't be. This one is tougher, because you can spend years learning the ins and outs of plumbing. Your best bet is to pick up a book or start watching YouTube videos, and then just get your hands dirty.

CHAPTER 3: SELF-SUFFICIENCY

How self-sufficient do you want to be?

Human beings, whether we want to admit it or not, are social creatures. We need to be around other people in order to be healthy and sane. The amount of time we spend around others, however, varies from person to person.

This is part of the reason people go into homesteading. Some of us just want to be left to our own devices. So we start scheming about ways to achieve self-sufficiency. For some, that means moving into the wilderness of Alaska and attempting to live off wild berries and fresh-from-the-river salmon. For others, that means being self-sufficient enough to reach out and create a like-minded group of people you can work with.

Both of these things are acceptable means of expressing self-sufficiency. And, with the advent of social media and near universal access to the Internet, it's possible to combine both a hermitical bent and an extensive support system.

Either way, self-sufficiency has always been and is a hallmark of homesteading. You'll have to determine exactly what your desired level of self-sufficiency is, and how to get there.

Again, this is one of those things it's better to ease into. You don't want to sell all your belongings and move out to the tundra thousands of miles from anyone and suddenly realize you just have to have a Starbucks nearby.

Separating needs from wants

Again, the best way to determine your wants and needs is to make a list.

Be brutally honest with yourself on this. When it comes to homesteading, wants and needs are fungible terms that can mean vastly different things to different people.

For example, my list doesn't include constant electricity, but it does include constant access to coffee and the Internet.

For my wife, her needs include constant electricity, as well as coffee and the Internet.

I don't require indoor plumbing, but I do require a method to water my garden and irrigate crops with ease.

Even when others, including members of your own family, consider unnecessary needs—like access to well-made frappes—to be frivolous, be aware that just because your needs don't align with anyone else's, it doesn't make them any less of a need.

Psychologically speaking, that frappe may be the difference between toughing out the hottest summer on record in a yurt without air conditioning and quitting and moving back in with your parents while you beg your former employer to rehire you for your previous data entry job.

Do not be ashamed of any of your needs!

As far as your wants, these are things that you'd love to have, but don't necessarily require to complete your homestead journey.

For example, maybe you need livestock and want angora rabbits. While you may wind up with space and time to raise chickens, you may find yourself in a situation where you just don't have the energy or time to feed, water, care for, and collect fleece from a high-maintenance fiber animal.

Since this is on your "want" list, you won't have to make a plan to purchase, house, and raise fiber rabbits only to find yourself in a position where you have to find some sucker, I mean good Samaritan, to take a giant, fluffy rabbit in.

Again, these are things that you don't have to be ashamed of.

Here's a brief glimpse of my wants and needs list.

Needs:

An irrigation system so I don't have to haul water or hoses

A well so said irrigation doesn't put me in arrears

Chickens for eggs, fertilizer, and to clear grassy areas

Goats for companionship and milk

Constant access to the Internet

Tons of people to chat with

A wheel hoe

An adult tricycle (to move materials and tools around on our homestead)

Wants:

A really cute mini truck I can paint turquoise and haul veggies to market in

A pedal-powered tractor

A stick welder

An emotional support chicken (it's a thing, I looked it up)

Now, while some of my wants and needs may not be everyone's cup of tea (an adult tricycle), I know by just glancing at the list how much time and money I'll have to spend procuring these things.

I'll also have to consider the wants and needs of my family (they tend to be sticklers about indoor plumbing, yet, oddly enough, are unconcerned about the adult tricycle—weird, I know).

Since we know what we need for our dream homestead, we can find property—or repurpose the property we already have—that meets those needs. We can also work on putting together a plan on how to fulfill our wants at a later time. For example, I'll want to make sure that whatever we decide to do, we have storage space, or the room to build storage space, for a pedal-powered tractor.

Again, this is just an easy and fun way to plan for your future while making sure you take care of yourself.

Necessities

While you're determining your needs and wants, you must deal with these categories: heating and cooling, electricity, water, shelter, and food. If you don't have these items on your list, you'll find yourself in a difficult position later on. You'll also need to come up with specific ways to handle each of these categories. Here's a breakdown:

Heating and cooling

One of the realities of life is that it will be hot in the summer and cold in the winter. You'll have to determine your personal comfort level with both these facts. Depending on where you live, you may be able to deal with the heat with strategically placed fans, shade, and iced tea. But that may require you to gather eight cords of firewood every year before winter rolls through. Other options for heating and cooling include earth tubes (geothermal drainage pipes to heat/cool air before venting into your home), wood heating, rocket stoves, thermal chimneys, passive solar, and so much more. Of course, there's always a good old-fashioned heat pump.

Electricity

Electricity can be provided via solar, geothermal, wind turbines, hydroelectric generators, and, of course, from the magic lines that go to most houses in America these days. Another option is to go off the grid.

Water

Municipal water has its issues: faulty testing, lowest bidder contracting, and so much more, but there's something to be said for turning on a tap and having water come out. That's not always guaranteed with a well or natural water source. Lines get clogged, pumps stop working, filters need replacing; it seems like if something can go wrong with a water system, it will.

If you're going to be doing a lot of farming, you'll need to put together some system to at least supplement your water usage. Collection of rainwater into barrels or a cistern, installing a shallow water well, or even digging a pond on your property

Going off the grid

Off-grid refers to doing without utilities provided by a co-op or municipality. Instead of hooking into the power grid, you use a solar, wind, or hydro-electric source, or any combination thereof. Or, you can even go without power. Instead of hooking into a municipal source of water, you can draw water from a well, spring, or even a natural source like a river or stream. All of these decisions have pros and cons.

Water sourced from your own property is going to be cheaper, a consideration for anyone using water to irrigate crops or providing water for large animals. Droughts and dry wells may impact you even harder when it happens. Your well or stream doesn't have large amounts of taxpayer dollars working to keep your pipes filled.

As far as electricity goes, it's a lot cheaper in the long term to produce your own. In the short term, however, setting up an off-grid electricity system is expensive and means putting a lot of infrastructure in place. And you may find yourself dependent on petroleum-powered generators for part of the year, when the sun and weather don't cooperate, or even losing power at inopportune times.

If you do decide to take that leap, there are even bigger rewards. It doesn't get more self-sufficient than producing your own water and power. With the increasing availability and reduced cost of off-grid systems, that decision doesn't have to lead to a life of deprivation.

Ensure the basic necessities, like water, shelter, and food, are on your list.

can all help alleviate the expense of your water bill, as well as provide an emergency source of water whenever needed.

Shelter

This is a fun one. Go ahead and find out now what kind of shelter you need for you and your family. I prefer tepees and yurts; my wife has a soft spot for log cabins. We both enjoy tiny houses. A friend of ours managed to purchase a farmhouse constructed in the 1800s. Dream big, but be aware that a specific type of home may be better served as a want than as a need.

While my dream home is a yurt in the Blue Ridge Mountains, it'd be difficult to raise a family of four as a mountain man in one of America's last great wildernesses. Which means we got a home with three bedrooms—the yurt just doesn't provide the privacy needs for a teenage son and a tween daughter.

Still, it's a lot of fun to sit and think and talk about, just try to be as realistic as possible to meet the needs of your family.

Food

I could live off dried beans, potatoes, and whatever happens to come up in my garden in a particular season. Supplemented with eggs from our chickens and dairy products from our goats, and as far as I'm concerned we're done.

However, nearly every member of my family is an avid baker, and it's hard to grow wheat on a small acreage. And they all love their meat. That means when we went looking for our homestead, I needed access to a location with a well-stocked grocery store, as I didn't want to start raising my own beef cattle. We did, however, also require a home in a town with at least one farmers' market and the ability to buy good, whole food for our family.

I know some people who can live off military surplus MREs and be perfectly happy. But when you're making your plan for self-sufficiency, make sure you understand exactly what you need and want to eat. Hopefully your homestead will be a forever home for you and yours, and if you need access to fast-food chicken nuggets to be happy, make sure that's part of your plan.

Electricity can be provided via solar, geothermal, wind turbines, hydroelectric generators, and from the magic lines that go to most houses in America.

CHAPTER 4: THE IMPORTANCE OF FOOD

Where does your food come from?

Food is one of the most important aspects of homesteading. In fact, I'd argue that food is often the starting point for most people who decide to begin homesteading.

Self-sufficiency starts with food.

Good health starts with food.

The American food system, to put it bluntly, can be a bit grotesque.

Most people do not know where their food comes from, what it goes through, and why food is prepared and presented to consumers the way it is.

So, where does your food come from?

Most of the fruits, vegetables, meat, eggs, and dairy in North America are produced on industrialized, large-scale farms. According to the United States Department of Agriculture, small family farms are, on average, 231 acres. One farmer feeds, on average, one hundred and fifty-five people.

Most of the crops grown on American farms are commodity crops: corn, soybeans, wheat, etc. The Food and Agriculture Organization of the United Nations reports that more than three hundred metric tons of corn were grown in America in 2013. In comparison, just over twelve metric tons of tomatoes were grown.

The only other vegetable crop to break the top twenty in tons of production that year was lettuce, with just over three metric tons (Note: A few high-value fruit crops, like grapes and apples, did make the list).

While the comparison between commodity crops and vegetable crops isn't exact—a head of lettuce weighs about one pound (sixteen ounces), while an ear of corn weighs twenty ounces—it does, in my opinion at least, speak to a lack of diversification in the modern industrial agriculture system. Most vegetables, more than half according to the California Department of Food and Agriculture, such as squash, beans, spinach, etc., consumed in the United States come from California. A sizeable portion—and a growing portion—comes from outside the United States.

Fun(?) fact: Less than 2 percent, according to a 2013 FDA report, of all imported food into the United States is not inspected by the USDA.

Most of the corn grown in America is used to feed livestock, according to the National Corn Growers Association, which is then sold all over the country.

Chances are, if you buy your food from a large supermarket or grocery store, that food has traveled thousands of miles to be there, according to a study

About 99 percent of corn grown in the United States is field corn, a commodity crop; the remaining 1 percent is the sweet corn you find in the grocery store.

put together by the Leopold Center for Sustainable Agriculture at Iowa State University.

Many supermarkets sell produce as a loss leader: a good portion of that produce is thrown away, spoiled before it's even sold. That means grocers don't make any money on fresh produce, but the material draws in customers who purchase other items, such as corn, wheat, and soy-based processed products, packed with preservatives, that take months, sometimes years, to go bad. The issues with a centralized system (in some cases, California is responsible for more than 90 percent of non-commodity vegetables consumed in America) are multifaceted.

A drought is sure to lead to a decrease in production over time, causing an increase in prices.

A disease outbreak can devastate a crop.

E. coli or a similar infection can originate in a single field, and then when the harvest is placed with the harvests from different fields for shipping the entire shipment becomes infected and winds up in grocery stores all over the country.

While disease outbreaks and crop failures can happen at any farm—even at the small family farm where they grow the tomatoes you buy at your local farmers'

Opinion Time

There are a lot of homesteaders, or those in the homesteading movement, who inadvertently vilify larger farmers. I truly believe that most of the time it's not intentional; however, it must be noted that I don't believe that anyone can criticize a farmer for the choices they make for their farms. Problems with the food system are systemic, and blaming an individual farmer for those problems is akin to blaming your mechanic when your car breaks down. It doesn't do any good and just leads to hard feelings all around.

Farming is hard, dangerous, difficult work, with very few financial certainties. That means farmers, however they choose to run their farms, are making choices that are best for them and their families. The best way to change the food system is through votes and dollars: vote for candidates, local or otherwise, who support the changes you want to see in your food system, and buy food that fits *your* criteria of how food should be grown.

market—the scale of large operations makes the impact exponentially bigger and harder to deal with.

A small farm operating on a local level can respond faster to issues like tainted supplies.

So, if you're interested in where your food comes from (and you are if you're reading this), it's probably best to find a farmer and talk to her. (More women are going into organic and small-scale agriculture than conventional ag, so you've got a solid chance that your farmer will be a woman.)

Ignore the labels—organic, natural, GMO-free, etc.—and just talk to your farmer. Discuss everything. How they grow their food, why they grow it that way, what their personal philosophy is—everything you can think of.

There are many reasons to grow your own food, and flavor is at the top of the list.

You're making a connection with a professional who provides one of the building blocks of your family's existence. This is a decision that most people don't even think about most of the time, but this is similar to choosing a doctor or accountant. Your farmer will impact many aspects of your life: your health, how much your food costs, family dinners.

A good relationship with your farmer will lead to tons of benefits, such as extras in your CSA box, first pick of new crops, and you'll learn more about your food in general.

Growing and producing your own food

Ultimately, if you're homesteading, you'll want to grow or produce your own food—even if you have the best farmer in the world.

There are many reasons to grow your own food: total control over growing methods, financial savings, the inherent joy of growing things. But be aware, it's nearly impossible to grow all of your own food. Even if you have all the space and time and money to create a completely closed system on your homestead, there are still things that you won't be able to grow, due to climate or soil conditions—coffee and chocolate spring to mind—or won't be worth growing.

That's okay. Since the beginning of agriculture, farmers have specialized out of necessity. Goat herders were too busy herding goats to grow grain, while grain growers were too busy growing grain to herd goats.

But that doesn't mean you shouldn't grow anything.

Even apartment dwellers can fill a windowsill with a little potted herb garden.

Just because you live in an apartment that doesn't mean you can't grow some of your own food.

City dwellers can rent rooftop space, participate in community gardens, etc.

Growing things will save you money and connect you to your food in a real way. You'll also increase the quality of your food in general. Using fresh herbs in your cooking is a revelation. Pulling radishes out of your garden is an epiphany. Making sauce with your own homegrown tomatoes is an ecstatic experience.

And knowing that you can grow your own food instills in you a confidence that is unmatched.

Grow food. Grow as much or as little as you can. Growing food is a cornerstone of homesteading. Even if you don't have a windowsill to put some pots on, you can sprout grains and beans in your kitchen. If you have a patio, plant a container garden. Use what you have to do what you can.

You can and should produce as much food as possible. You can grow veggies nearly anywhere, but don't discount other options.

Look into meat, honey, egg, and milk production. It may be a step too far for you, raising animals to slaughter and butcher at home, but you can always get a couple chickens to produce eggs, bees to make honey, or goats to milk.

All of those things are incredible calorie producers, which means you'll get a lot of energy out of the food produced by those animals.

While you're at it, don't discount raising animals for meat. There is a definite ethical and moral consideration at play, and it could be argued that everyone should know what it's like to slaughter and butcher their own meat. It provides a person with perspective on exactly what goes into putting meat on her plate. Again, even if you don't have space, there are options, such as backyard chicken coops and rabbit hutches. A lot of animals can be raised in a space just slightly bigger than a small bedroom. I even read a blog post on how you can raise dozens of edible snails in a plastic bin you can keep in a closet.

One of the unusual things—other than the snails— you might want to consider is mushrooms. Mushrooms can be grown in small spaces, have tons and tons of nutritional value, and so much more. Oyster mushrooms, much like snails, can be grown inside plastic bins, which are stackable.

You should think of these concepts in a completely different way. Yeasts can be grown in sourdough starter. Bacteria can be grown in kombucha starters (also known as SCOBY—symbiotic colony of bacteria and yeast), and in sauerkraut. Essentially, this entire action, the act of growing and producing the food you need, is an action of marshalling the life that is around you in a symbiotic relationship that provides nutrition for you and your family, while encouraging you to experiment and explore that relationship.

All of these items, from the macro (chickens and veggies and mushrooms) to the micro (bacteria, yeasts, and more) are acts of exploration and discovery, and

Organic. Local. GMO. Cage-Free. Natural. These labels can be confusing until you know what they mean.

only by exploring and exploiting that can a homesteader truly achieve self-sufficiency.

What do all those labels mean?

While you're shopping for food, you should be knowledgeable about the terms you'll start to see on the labels of many products. Below are definitions for some of the more common terms.

Organic—Organic usually refers to a holistic growing system that attempts to create a system of food production that is sustainable. Organic does not mean chemical, and it does not mean "no pesticides." Organic farmers use pesticides, many of which are derived from plants and processes considered safe and minimally damaging to the environment. Organic certification is provided by many different organizations with a fairly diverse set of standards. Which means the standard used by the United States Department of Agriculture might not be the same as the California Certified Organic Farmers standards. So, if organic certification is important to you, you'll have to learn what that means.

GMO—Or, genetically modified organisms. These refer to the process of genetically modifying a living thing, mostly a plant, by using a laboratory process that transfers a sequence of DNA from one living thing to another. The process is controversial, as many opponents point out the consequences of such actions can be unforeseen. Proponents of GMOs state the modifications aren't any more harmful than other agricultural practices. Most GMOs aren't currently labeled, so you'll have to look for "non-GMO" labeling if it's important to you. Be aware, however, that some companies don't advertise their GMO status, one way or the other.

Local—Local food doesn't have an agreed upon meaning. Some places consider local food to be within a city/county boundary, while others set a mileage radius from a specific location. Some farmers' markets state that local is as far as a truck can drive in a single twenty-four-hour period, which can be more than seven hundred miles. Which means that for an New York City farmers' market, "local" food can come from as far away as Kentucky.

Natural—According to the Food and Drug Administration, the term "natural" on food has no definition. In America, however, the term is used primarily for foods that don't contact synthetic additives or artificial coloring. But, ultimately, the term does not indicate much about the origin of the food, how it's made, or what it's made of.

Cage-free—Another term without distinction in the United States Department of Agriculture. However, it is defined by the American Humane Association and covers a list of requirements, including ventilation and space. The space required by the American Humane Association is one and a half square feet, about the size of two sheets of notebook paper.

2 | Section Two: Farming

CHAPTER 5: WHERE DO I GET LAND?

Renting versus owning

When we first decided to start homesteading, before we even knew that what we were doing was called homesteading, we went searching for land. At the time, we'd already defaulted on one mortgage—like many American families, the 2008 financial meltdown hit us hard: we were laid off and wound up broke.

In fact, that was probably the catalyst for our homesteading journey; we never wanted to be in a position of such insecurity again.

So we started looking for new ways of doing things. At first we thought running a full-time traditional goat dairy was the way to go.

But as we researched and found out how much property was needed to keep a herd of dairy goats, as well as the cost of grain and feed, it just wasn't feasible—a quarter million dollars spread out over a thirty-year period meant that our dream project wouldn't be paid off until we were both in our sixties.

So, we tried strange and unusual ideas. We moved our family of four into an eighty-square-foot tiny home.

We tried different things with family members, leasing land, rental agreements, etc. And while we wound up buying about four acres in North Carolina with a ten-year note, we learned a lot in the process.

While bouncing around from place to place, we found out that homesteading and renting were not mutually exclusive. All the ideas we tried, from our tiny home to raising dairy goats on family land, actually worked, to some extent. Even when we found ourselves in a situation where we couldn't plant a garden in our backyard, we leased land specifically to farm.

And we found out something that was amazing: leasing farmland is certainly cheaper than buying it.

Farmland in North Carolina leases for about $250 a year! And that's if you go looking for farmland. If you look for odd little bits of land—vacant lots in your city, an elderly neighbor's backyard, or a patch of growing space in your community garden—you'll actually pay a bit less for the space and sometimes be allowed use of the space for free.

Leasing land does have its drawbacks: Sometimes you won't be able to build structures on the property, and you may have limited access to water and other amenities, but the whole process is a lot cheaper than buying forty acres in the middle of your rural dream.

Don't take buying off the table, however; just try to be a bit more picky when making that commitment. Look for contract for deed offers; the interest rates are usually lower, and, therefore, the mortgage payment doesn't last as long.

If your living situation doesn't provide you with a place to grow your own fruits and vegetables, a community garden may be the perfect solution.

Empty, out-of-the-way lots without water or electricity can usually be had for just a few thousand dollars an acre, making an off-grid homestead well within the reach of most people.

Ultimately, though, beginning with a bloom-where-you're-planted attitude goes a long way and lets you get started.

Even if you do find yourself in a situation where you're moving every few years—think military families—then you might feel like there's no point in doing anything until you're in a "stable" situation. However, that is incorrect thinking. Point of fact, there's no such thing as a stable situation. In the meantime, while you're waiting for "stable," you can be feeding your family, learning new skills, and finding out about your limits. All of which are a lot better to understand before you spend a lot of money on your dream property only to find out what you thought homesteading was isn't for you.

Either way, whatever you decide to do, buy or rent, there's a few options for farming and homesteading you should be aware of.

Yard farming

Experts estimate there are about forty million acres of grass in the United States alone. That means just about everyone with a lawn is spending hundreds, maybe thousands of dollars a year on raising a crop: grass. With some effort, the average suburban yard can become a working farm in less than a year's time. And it's truly astounding how much can be produced.

A farm in California, the Urban Homestead, produces three tons of food each year just outside of Los Angeles on about one-tenth of an acre in yard space.

Instead of raising grass, turn your yard into a place where you can grow herbs and vegetables.

You might not want to grow that much food at any given time (three tons is a lot of food), but it illustrates that with the proper planning and work, there's no limit on what can be done in an average American yard.

Yard farming also has some great advantages: your water is right there where you need it; you won't have to purchase a lot of expensive equipment, like tractors and harvesters; and you're likely to already have a yard.

If you're interested in farming your yard, you'll need at least a few resources. Check out farm-a-yard.com and also check out Chris McLaughlin's book, *Vertical Vegetable Gardening*. Both resources are excellent for maximizing space in a small yard farm.

Basic steps to convert your yard into a farm

1. Get rid of the grass. You'll either have to remove the sod, mulch your grass, or install raised beds. Look for areas in your yard that get at least six hours of sun a day. Also, be aware, the position of the sun changes based on the season, so a spot in your yard that looks bright and sunny during the winter may be in the shade for much of the day in the summer. Also, get your utilities marked if you're going to be tilling or digging. You don't want to destroy your electrical lines to plant beans.
2. Get your soil right. Once growing space has been cleared off, you need to get the soil tested. Testing services are often provided through your local USDA Cooperative Extension Agency, as well as many universities for a minimal cost. Soil testing can also be done through private companies if you're interested in test results coming from an alternative growing perspective, say, organic or biodynamic for example. Once you get the results of your soil test, you'll know which amendments to add to the soil to get it ready for planting.
3. While time-consuming, soil testing pays off big in the long term. The weeks you wait for your soil test results to come back will return to you in increased harvests and money saved, as you won't overspend by putting amendments into your soil that you don't need.
4. Mulch everything. More than anything else, mulching is the single biggest thing you can do to make your yard farm a success. Mulching promotes microbial growth in the soil, prevents erosion, and protects earthworms. Mulch can be expensive, but finding mulching material, if planned for, can be easy. Shredded newspaper makes good mulch, so do tree leaves from the fall, and wood chips. Some landfills sell mulch at a low price, and some electric companies will drop loads of ground tree limbs off at your home for your use.
5. Don't forget animals. A few chickens, rabbits, pigeons, or quail can make a

big difference in your homestead. From animals you can get meat and eggs, as well as manure, which is more helpful than anything else. Manure provides many of the nutrient requirements for plants, while grazing animals help control weed and grass growth.

6. Make sure you know the law. If the zoning regulations or homeowners' association rules don't allow the basics of yard farming, then you can lose a lot of money going against those regulations. I'm not saying you shouldn't work to change them, but you don't want to spend two years putting in a yard farm only to have a cease-and-desist letter sent your way and having all your hard work go to waste.

811—Call before you dig

Before any project that involves digging in your yard, such as preparing the land for a garden, fence installation, or planting a tree, a call to 811, Dig Safe, is the first step. Every state has its own call center. When you call, be prepared to give them the details about your project and your location. Dig Safe will notify the utility companies, which will send someone out to your property to mark buried lines so you do not hit them once you start digging.

Micro farms

Yard farms, and urban farms for that matter, are a type of micro farm. In my experience, yard farms tend to be smaller, whereas micro farms tend to be less than five acres in size, slightly bigger than their smaller cousins. Technically, there's no standard definition for a micro farm, as "small" family farms are, on average, two hundred and thirty-one acres, according to a United States Department of Agriculture 2012 census.

All of the tips on developing a yard farm apply, but there are two big differences: infrastructure and equipment. Homesteads bigger than one acre will need just a touch more in the way of equipment than yard-size operations, particularly in the form of a small power tiller, broadfork or wheel hoe. Extra equipment also means extra infrastructure, particularly in the form of shelter for your tools. Micro farms might also require extra plumbing to provide irrigation nearer to crops. All of these things are potential operation wrecking costs to a homestead. It also must be noted that more land doesn't always mean more produce.

Unlike other production operations, the economy of scale rule doesn't always apply to farms, which cut into cost-per-calorie and cost-per-unit calculations. Therefore, a good rule of thumb with homesteading is to aim for what you *need* instead of what you think you *want.*

Urban farms

Urban farms aren't determined by size, but by location. Any farm within an urban area is considered an urban farm. While these operations can be micro farms or yard farms, it seems one defining feature of an urban farm is a greenhouse.

One of the challenges of urban farming is due to the microclimates within city limits. These microclimates are caused by shade from tall buildings and the canyon effect these same tall buildings create, moving heat and wind around in unpredictable ways. Greenhouses allow urban farmers a way to offset these unpredictable climate patterns and provide a way to extend the growing season to make up for smaller growing spaces. Greenhouses, however, aren't a necessity in urban farms and can be expensive to build and install, so if you choose to utilize one, be sure to not overextend yourself financially.

The benefits to homesteading in urban areas include the availability of markets and customers—and supplies—if you have a professional operation. And while many people crave the solitude of the country and wide open spaces, homesteading and farming can be lonely occupations. Being in an urban area puts you in contact with people, so you can build a community of like-minded individuals who can help with crop mobs, buy your produce, and maybe even make you feel a little less crazy when you decide to quit your job and start an urban farm.

Definitions

A **wheel hoe** is a push hoe with one or two wheels. Most models come with two handles. The hoe attached to the device is often removable and interchangeable with discs, sweeps, and plows. The device works really well for well-kept, loose soil and is great for operations desiring a small carbon footprint.

A **broadfork** is a two-handed garden fork that works great for loosening and aerating soil. With the right planning and work, a broadfork can replace a power tiller.

A **tiller**, either powered or unpowered, stirs earth and soil by digging and overturning it. Most power tillers overturn the earth. Although some also cultivate it—a form of secondary tillage. Cultivation, for most intents and purposes, tends to occur after the first tillage and breaks the soil into finer particulates. While walk-behind power tillers are the most well-known, tilling can be done with a shovel, a broadfork, a large digging hoe, or a manual twist-style tiller.

Power, handheld (think claw cultivator), and rotary **cultivators** are used to break the soil medium into finer particles. Cultivation also plays an important role in weed suppression during the season, as cultivation allows gardeners the ability to remove weed biomass from an area.

Hula hoes feature a circular, open blade at the end that resembles the stirrup on a horse's saddle. They're also called action hoes, because of the movement the blade produces due to the swivel on the neck of the hoe. These are, hands down, my favorite tool to use. They come in sizes up to seven inches wide and make weeding a lot less of a chore. While they work best in well cultivated soils, they're also tough enough that a backyard gardener can use them to remove patches of sod when necessary.

CHAPTER 6: HOW TO FARM

The basics

When we started on this journey, the only thing I knew about farming was that smarter people than I put seeds into the ground and got food out. Some of the really smart ones fed their families and made money with that talent.

While my saintly Granny Jones could feed an army out of her garden, I learned nearly nothing from her, as I was a hard-headed child in a family of hard-headed children.

So, our journey began with a lot of ignorance coupled with a lot of desire. We knew that food grown closer to home tasted better and was fresher, but we had no idea how to do it ourselves.

Being the book-learning geeks that were are, we immediately started reading and trying things out. We started with books like *Square Foot Gardening*, *Lasagna Gardening*, *Vertical Vegetable Gardening,* and many more.

We started attending classes at the local USDA Cooperative Extension office and searching for every tidbit of information we could find on websites like Rodale's Organic Life and all the information from permies.com. We bought a copy of every farmers' almanac we could find.

That massive information dump is one of the things that led to forming our magazine, *From Scratch*.

After getting all that information packed into our tiny little heads, we started gardening. Of course we were going to do it organically. And, we failed.

We had no idea we'd have to fight weeds and insects as much as we did. We had the barest understanding of how the growing seasons worked. We barely understood soil fertility and microbial activity and had very little understanding of how to read or use a soil test.

We tried everything: building raised beds, plasticulture, a bizarre form of no-till, and so many silly little ideas. Some worked, most didn't.

While it was discouraging, every failure led to new successes. We learned how to control weeds with mulch and discovered new types of tools—the hula hoe with a swivel head is life changing. We learned how to combat insects with organic materials, essential oils, and environmental controls. We figured out how to read a soil test result. We learned how to deal with late frosts and early heat.

Finally, after about three years of gardening and trying, we had a semi-successful harvest. We actually broke even! While that may not seem like a big deal, it is. Most farm operations take about five years to break even, sometimes even longer.

After several seasons of gardening, we had a semi-successful harvest. Along the way, we learned about frost, soil tests, insect and weed control, and so much more.

The other thing we learned—and this is the real success—is how we'll never stop learning.

Insects and pathogens sometimes have life cycles that stretch over years, which means some years you won't have an insect problem, and the next year a wretched creature comes out of the ground and destroys a crop.

Even without climate change, weather patterns are unpredictable and fickle. The weather reports are mostly right, within limitations, and you'll learn to panic when your meteorologist predicts temperatures above freezing, because even if it's five degrees above freezing, you know from previous experience that you'll still wake up to a frost in the morning.

Farming isn't a scholarly process; it's a verb, which means you learn by doing. That isn't to discount book learning. Everyone needs a starting point, and if you don't have some sort of knowledgeable foundation, it's going to make your failure rate higher. Failures can lead to frustration, and frustration can lead to quitting.

The way to combat failure, however, is to embrace it head on. Accept that you'll always be a student of farming/gardening and enjoy the learning process. Every day you spend in your fields or in your garden, even when you have to till everything under and start over, is a good day as long as you learn to enjoy the rhythms and movements of nature.

So, how do you start farming vegetables? Best way to go, as far as I'm concerned, is to find a book that looks interesting and get started.

To help you pick and choose a method, below are some basic farming methods explained. There are a lot of other methods that fall under these broad categories (and a couple of them are subcategories already), so you'll need to determine which one works best for you.

Conventional farming

Conventional farming is exactly that: what is conventionally done. While this term is often bandied about, it doesn't mean there's a standardized method of farming done on every farm. Conventional farming is also known as industrial farming.

Usually, conventional farmers use every technology they can to get a crop to market: chemical pesticides and herbicides, industrially produced fertilizers, heavy machinery—really, anything to get a crop to market. Philosophically this method is pragmatic at its heart. Nothing matters but dealing with a crop and its issues in the here and now.

Conventional farming practices are often associated with large farms, but they're also a big piece of many small gardens. My saintly Granny swore by Miracle-Gro and Sevin dust. Many backyard gardeners and farmers spend lots of money and time on premixed fertilizers and sprays to prevent disease and insect infestation. This model of farming and gardening is very popular, because for the most part it works. However, it isn't sustainable over the long term.

By depending so heavily on outside inputs, like chemical fertilizers, the garden's soil is virtually ignored, which means as soon as the inputs aren't provided, the soil stops producing. Since many of these chemicals are derived from petroleum byproducts—for example, ammonia is produced with hydrogen from oil—fluctuations in the cost of oil leads to wildly varying prices, making it hard for many of these operations to budget properly. For example, in 2016, ammonia prices dropped 20 percent from January to July, and in August they climbed nearly back up to the original price.

Since much of this farming method is petroleum based, many ecology-minded critics worry about the ability to continue farming this way in the coming decades, as a major depletion of oil resources would drastically increase costs. Conventional farming and gardening is unsustainable at this time. However, I don't want anyone to assume that this is a criticism toward

Conventional farmers will utilize every available technology they can in order to get their crop to market.

Food-grade diatomaceous earth can be used as pesticide and has a minimal impact on soil health; however, it will also kill off beneficial insects.

farmers and gardeners who use these methods. Conventional methods are often the only way these producers have to operate, considering the demands of farming in the twenty-first century. And many of the organic methods we'll discuss in this book aren't suited for large-scale application over hundreds of acres of production, as the infrastructure required for these things aren't incorporated into the modern American agricultural system.

Organic farming

Organic farming is the diametric opposite of conventional agriculture in philosophy, but not always in methodology. While organic producers work hard to avoid industrially produced pesticides and herbicides, they aren't always able to. The primary goal of organic farming is not avoiding these things, but instead acting in a manner that is sustainable and holistic. That means the standards of care for an organic operation are a bit different. The overall effectiveness of a pesticide has less impact on deciding to use it than the effect the pesticide has on the local ecology. To give you an example, one of my favorite pesticides to use is a product called food-grade diatomaceous earth (DE). It's completely safe for humans and animals; it's even used by some as a supplement.

There's some disagreement on how DE works. It either scratches the carapace on an insect, leading to tissue failure, or it dehydrates an insect, leading to death. Either way, it's a very effective pesticide with no harmful side effects to humans or animals, and it has a minimal impact on soil health over the long term. In fact, it may actually improve soil health by increasing calcium levels slightly with long-term use.

I rarely ever use it.

Why? Because it's an indiscriminate killer. Lisa Rayburn, an extension agent in Onslow County, North Carolina, explained it to me, and I'm thankful for her expertise (Pro Tip: Find smarter people than you to learn from. For me, it's easy, as nearly everyone is smarter than me, but if not, then learn from Lisa. She's an excellent soil scientist with years of knowledge and experience under her belt.).

Since DE damages any insect with an exoskeleton, which is just about all of them, it also kills insects that I like: bees, ladybugs, and assassin bugs. Lisa explained ways I could minimize the damage, but since bees have it hard enough, I ultimately decided that it wasn't worth using, except in very specific situations, where I know beneficial insects won't come in contact with it.

For me, and for most other organic farmers and gardeners, that situation is a perfect example of what organic methods are: a measured and carefully considered response to agricultural challenges that encompasses the needs of the farmer, the safety of the consumer, and the health of the ecology itself. While there are many different methods of achieving those things, organic producers all share that holistic approach.

I encourage everyone to bear that in mind during their homesteading journey. Not everyone will agree with you on how you achieve your homestead, but chances are, if they're interested in the movement, they will support the same philosophies and ideas. So don't use litmus tests!

Permaculture

Permaculture is an offshoot of organic principles and views agricultural systems even more holistically, if possible. Just as much philosophy as agriculture, permaculture advocates expand the concept of farming into a reflection on human existence in a lot of ways. A pioneer of permaculture, Masanobu Fukuoka, called the movement "do-nothing farming." Permaculturists thrive on three main principles: care for the earth, care for the people, and return of surplus. Essentially, the idea is that caring for the earth leads to greater yields with less work and expense, which allows permaculturists to care for the people, who must serve as stewards for the earth. The people return the surplus of energy and materials back into the earth by utilizing compost and dedicating time saved to increasing their stewardship.

Permaculturists thirve on three main principles: care for the earth, care for the people, and return of surplus.

Done properly, it's a beautiful system. But the system takes a lot of time and energy and doesn't necessarily put the desires of the individual gardener or farmer front and center. For example, if you want to grow a crop not normally associated with your climate or region, permaculture and its advocates would strongly advise against that, as doing so would increase the amount of inputs and energy required, which could be used for other cops and systems that would produce more food with less effort.

Another drawback of permaculture is the high initial investment. Food forests, a key component in many permaculture operations, usually won't provide a high level of return on investment for about three years, the time it takes for trees to produce a good fruit harvest. This actually makes the method better suited for small operators; it's a lot cheaper to keep one acre going for three years while waiting on a harvest to come to fruition than it is to keep two hundred acres running for three years.

To learn more about permaculture and its methods, check out Fukuoka's book, *The One-Straw Revolution*, Bill Mollison's *Introduction to Permaculture* and David Holmgren's *Permaculture: Principles and Pathways Beyond Sustainability*. Also check out anything by Joel Salatin. Salatin's work with holistic pasture management and meat production at his operation, Polyface Farms, has made him a sustainable agriculture star.

Biodynamics

Biodynamics is, in my opinion, one of the most fun and controversial methods of organic agriculture in use today. It's also probably the oldest movement in organic agriculture. Derided by critics as witchcraft, pseudoscience, and snake oil, the movement emphasizes a spiritual approach to agriculture and farming.

Some of its methods share similarities with notions of animism and sympathetic magic, which are often associated with folkways and tribal cultures.

Surprisingly enough, however, it appears to work. In the nineties, the Egyptian Ministry of Agriculture converted all of the country's commercial cotton acreage over to biodynamic production, reporting a 30 percent increase in yields. Biodynamic vintners report increased yields and better wine.

The whole movement has a storied history. Rudolf Steiner, a philosopher, held a series of lectures on agriculture in the early 1920s. The lectures were later expanded upon to create the biodynamic movement and is considered the beginning of the modern organic agriculture movement.

Definition

A **food forest** is an orchard designed to produce food for animals and people, and over a length of time integrating more than just fruit and nut trees. For example, animals will sometimes graze in between trees in an orchard, fertilizing the ground with their droppings and reducing competition for resources. Leafy greens or shade-loving vegetables will be grown under trees, increasing the efficiency in how the space is used. Younger trees can serve as supports for climbing vegetables, such as beans, squash, and cucumbers.

Its critics accuse the movement of being a hoax, and skeptics claim the movement's results can be replicated with any form of organic agriculture. Its methods, to be fair, can be downright weird. There's strange combinations of materials, including cow horns, ground quartz, and stag's bile. And crops are planted according to a calendar that takes into account the movement of the planets, moon, and constellations, giving it an astrological feel. But, like I said, it's pretty neat, kind of fun, and a surprisingly economical method of organic agriculture. If you're interested, pick up a copy of the Stella Natura, the biodynamic planting calendar, *Biodynamic Gardening*, a DK book; or *A Biodynamic Manual: Practical Instructions for Farmers and Gardeners* by Pierre Masson.

Weed Control

Weeds are, hands down, the biggest hurdle many beginning farmers and gardeners will face. Not having a plan for weed control is like having a weed growing plan; you're either fighting them or encouraging them. Here are some, mostly organic, weed control methods.

Plastic—A heavy-gauge plastic will block out sunlight, killing weeds. You can use a heavy reusable tarp or any dark color plastic to get the job done. Some farmers will put the plastic down and then cut holes in to plant seeds and transplants. It's a relatively good method of weed control, but the plastic will break down due to UV radiation. It can also be difficult to remove at the end of the season, which means you could wind up with bits of plastic in your soil. This prevents many pure-minded organic gardeners and farmers from using it. The plastic also prevents water from reaching the soil, so using a drip irrigation system is necessary.

Solarizing—Solarization involves using sheets of clear plastic to cover patches of earth. Sunlight penetrates the plastic and creates a greenhouse effect underneath it. The soil and the weeds around it are cooked, killing weeds and many weed seeds that live in the soil, which prevents new weeds from growing after the plastic is removed. This method works best during the hottest summer months. Again, a heavy-gauge plastic is required, which will break down under UV light. The plastic needs to be tight against the soil and sealed at the edges. Garden staples work well to keep it tight to the ground. To seal the edges, bury them in a couple of inches of soil. Let it sit for at least a few weeks.

Cultivation—Tilling weeds under as they come up, either with a tiller, hoe, or cultivator, will eliminate most weeds. It is a lot of work, however, and a time commitment. It must be done often, at least every three days, in order to be effective.

CHAPTER 7: ANIMAL HUSBANDRY

The basics

For me, a homestead isn't a homestead without animals.

When we first started homesteading, we wanted to get goats. Goats are the most noble animal ever put on earth. They've got all the personality and intelligence of dogs, they're smarter than sheep, and they enjoy the company of humans and other animals.

But we didn't get goats, at least not at first. We started with some chickens, later moved to ducks (dastardly creatures), rabbits, guinea fowl, and eventually goats.

One of the goats we got was a dwarf Nubian mix, which meant she was small. She was sickly. She came down with a respiratory infection right after I brought her home. She was also tough. She'd get into fights with the Alpine dairy goats we had, animals easily twice her size.

She wasn't very well mannered. It took her weeks to get used to us. We milked her for a while, and she bucked and kicked the whole time.

Finally, months later, we established a bit of a relationship. We understood each other. We were starting to get along. She knew I meant food and water and no harm. Occasionally she would get her head stuck in the fence while trying to reach for a bit of vegetation just outside the enclosure. Then she wouldn't be able to pull her head back through the wire because her hooked horns would get stuck. Then she'd yell for me to come and get her and patiently wait while I got her untangled.

Every night I'd go and check on her and the other goats to make sure they were bedded down and safe.

When winter came, I put straw inside the little goat house I built for her and her herd mates to sleep in and stay warm. The weather called for snow overnight. I spent most of the evening standing in the sleet and rain trying to get the stovepipe on the wood heater fixed so I could build a fire and stay warm through the winter storm the weather service had warned us about.

After the rain froze in my hair and beard, I warmed up a bit by the heater, and then I went to check on the chickens and rabbits and goats.

They were all fine.

I went to bed, warm and happy about getting so much done that day. I woke up, had a cup of coffee, and then went to check on everything. The rabbits were fine, the chickens were fine, the horrible ducks were fine. The goats were fine, except for one little girl goat that was missing. I looked all through the enclosure, calling for her. When she heard me, she started bleating. She was lying next

Goats have all the personality and intelligence of dogs, they're smarter than sheep, and they enjoy the company of humans and other animals.

to the fence, her head twisted back at an awkward angle and unable to move. She was cold to the touch. I scooped her up and took her inside the house. I gave her a hot bath to warm her up, got her some food and water, and started making phone calls. Vets and vet techs were of little use, as most of them in the area knew very little about goats. I researched and talked to other goat keepers and took a trip to the feed and seed store to get medicine for the little girl goat. Most people suggested putting her down. I didn't want to do that.

I spent three days putting vitamin supplements and medicine in the girl, kept her warm and comfortable, and made sure she had plenty of water. Three days after I found her out in the cold, I was sitting in the garage with her, petting her. She laid her head on my lap, sighed a bit, and died.

We learned a lot from keeping animals, a whole lot. We learned how to keep animals clean and safe, we learned how to feed animals without going broke, we learned how to feed ourselves with the help of animals, and how to incorporate those animals into growing vegetables and more. We learned how to inject goats, dress the wounds of chickens, and which sulfa drugs work on intestinal parasites.

We saw baby chicks being born, and we watched little girl goats die.

Having animals on your homestead makes it a little more fun, and a lot more meaningful. You get to be partners with your animals, even the animals you kill and eat.

You and your animals will work together to feed each other and care for each other. You'll provide food and health care and shelter for your animals. They'll provide food, entertainment, companionship, and even income for you.

Meat, milk, eggs, fish, fiber, and more

Let's say you don't particularly like animals. Why should you keep them?

Animals provide meat, milk, fiber, and more. They add value—and calories—to any homesteading operation, and they help make huge strides for any plan toward self-sustainability. Here are some examples of animals you can raise and what you can do with them.

Meat

Meat animals include chickens, quail, rabbits, ducks, cows, goats (yes, goats), pigs, sheep, basically any animal you can think of. For smaller homesteads, I suggest

When people think of milk, cows usually come to mind; however, homesteaders should consider other options, such as sheep and goats.

Meat breeds

Rabbits—The most popular breeds for meat are Californian and New Zealand. These breeds can weigh up to ten pounds each. Flemish Giants are sometimes raised for meat, and are the largest rabbit breed in the world. They can reach weights of twenty-two pounds. They aren't that popular for meat rabbits though, as they take longer to raise to butcher weights.

Chickens—Popular breeds for meat chickens are the Cornish Cross and the Delaware. While broilers are often raised for seven to nine weeks and then slaughtered, I actually prefer to raise birds through their egg-laying years, usually three years, and then processing the birds for the meat. You'll get a lot more value per bird this way. Some sources say the meat is healthier, but it's also tougher; if you're raising meat birds for market, many consumers will not know what to do with the vastly different meat texture found in older birds.

Goats—Hands down, the most popular meat goat is the Boer. These monsters can weigh up to three hundred pounds. Another option for smaller homesteads is Pygmy goats. This tiny breed originates in North Africa and was originally bred for meat. They're built like tiny steers: all muscle.

Ducks—The most popular meat breed of duck is the Pekin duck. This is the stereotypical white duck with a bright orange or yellow beak. Ducks have become increasingly popular as farm and homestead animals. Bacon made from duck breast is supposedly better than its porcine counterpart.

starting with chickens and rabbits. Both animals are relatively cheap to care for and don't require much space. Chickens can produce eggs for the first three years of their lives and then be slaughtered, which means you get double duty. Rabbits are incredibly efficient—for the same amount of feed and effort put into raising a pound of, say, beef, you can raise more than five pounds of rabbit meat.

Goats are a good option for larger homesteads, as meat goats can be raised on grass. Kunekune pigs, a heritage breed from the Pacific Islands, does well on smallish homesteads and can be raised to market weight on grass alone. Quail can be produced in a really small space, but the amount of meat you get from them is limited, making these tiny and tasty birds better suited for market and restaurant sales than for home meat production.

Milk

Milk provides homesteaders with a unique opportunity. It provides lots of calories: a gallon of goat's milk provides nearly three thousand calories. It can also provide lots of income; homemade cheeses and other products can be sold at a premium at farmers' markets and at many restaurants.

Cow's milk is the most popular version of dairy from farms, but homesteaders have other options that are more practical. Alpine and Nubian goats are known for their milk production. Sheep can be milked. The East Friesian is the most popular sheep dairy breed, and the milk from these animals can be used to produce Roquefort, ricotta, pecorino, and Romano cheese. Some homesteaders and farmers have even reported success in milking camels. Camel milk is highly prized in the Middle East and is recommended as a remedy for allergy issues by some practitioners of naturopathy. Homesteaders, even smaller ones, can also invest in cows. Miniature Jerseys and Holsteins produce less milk than their larger, full-figured cousins, but they can be raised on less space with lower food costs. In addition, homesteaders might look into something like the Shetland Cattle, which are naturally smaller. These cows can be raised for meat and milk production, making them an excellent choice.

Eggs

Just about every egg produced can be eaten. Chicken eggs are obviously the most popular. This is also a great way to bring in extra income to help feed your birds. While chicken eggs are the most popular, duck eggs are rapidly becoming popular among foodies all over the country. Duck eggs fetch a pretty penny, too. Some sources sell duck eggs for five dollars each, which means duck eggs can sell for up to sixty dollars a dozen!

A word of advice for those who sell eggs: don't sell the eggs for less than it takes to produce them. I did the math recently, and with layer pellets costing fifteen dollars for twenty pounds (probably the most common way to feed chickens) a dozen eggs is three dollars and sixty-three cents in feed costs alone. That doesn't include the original cost of the birds, or the cost to raise them to maturity. Most egg sellers I know only charge two to three dollars a dozen, which means they're probably losing money. No matter what your reasons for raising chickens, if you lose money doing it, you probably won't do it for long.

Fish

Fish are unique and valuable products for the homestead, but they have unique requirements compared to other homestead animals, namely a pond or aquarium. Fish can provide excellent fertilizer for your garden or farm operation, they're great at controlling pests— many fish thrive on mosquito larvae—and they are excellent sources of protein. Additionally, and tragically, depleted fisheries around the world mean that farmed fish can fetch a good price at markets.

Koi fish are one of the most profitable types of fish. A well-graded koi fish can fetch up to five thousand dollars. And while small operations are rare, I'm of the opinion that fish eggs, or roe (also known as caviar), are a potential source of revenue for a homesteading operation.

One of the most commonly farmed fish is tilapia, which are surprisingly versatile for homesteading operations. Systems made with fifty-five-gallon barrels, readily available just about anywhere for a small price, can produce forty to fifty fish a year, while taking up less than ten square feet of space.

Fish can provide fertilizer for your garden, they help control pests, and they are an excellent source of protein.

Fiber

I'm an avid crocheter, and I believe that spending lots of time participating in the fiber arts is good for you and your homestead. It's meditative, it gives you something "productive" to do in the wintertime, and it's a potential revenue source. Many animals produce fiber, be it sheep, goats, alpacas, or even rabbits.

Sheep, goats, and alpacas might not be feasible for fiber sources for many homesteaders, as they require a bit of specific knowledge; shearing is an often unrecognized art form. And while I encourage everyone to learn new things, it may not be possible to learn in your area. Fiber from angora rabbits, however, can be harvested with a blow-dryer and a tarp. The fleece from Angora rabbits is valued by knitters and crocheters. A single rabbit can produce nearly one pound of fiber a

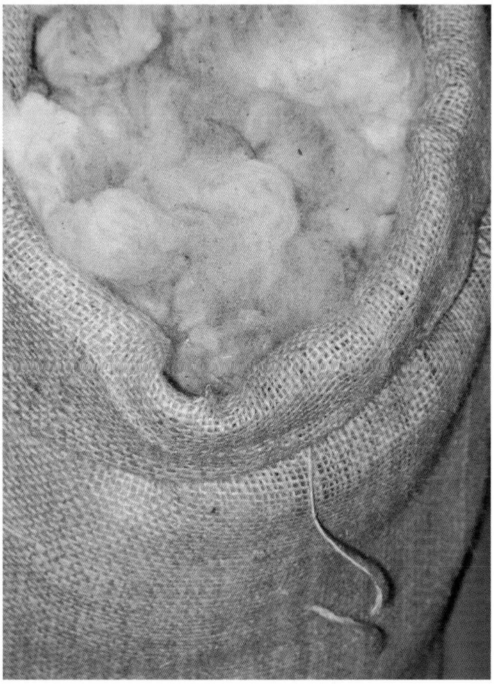

Many animals producer fiber and provides an option for earning money from the animals without harvesting them for meat.

Products like honeycomb can provide an alternative revenue stream for those who keep bees.

year. And like meat rabbits, they're easy to raise in small spaces.

While fiber production and sales is often overlooked by homesteaders, fiber production does provide a method to raise animals without harvesting them for meat, which some homesteaders aren't entirely comfortable with. To learn more about rabbits, check out Chris McLaughlin's book, *Hobby Farms: Rabbits: Small-Scale Rabbit Keeping*. It's an excellent resource on all things rabbits.

Bees, bugs, and worms

Insects and invertebrates, while not normally considered livestock by some, are definitely a way to expand your homestead. In fact, I'd encourage every homesteader to start an earthworm bed before doing any other animal production. Worms provide you with fertilizer that is perfect for your garden. They also serve as an alternative protein source for chickens or fish, if you choose to raise those animals at a later date. It's also possible to sell earthworms to local fishermen for some extra cash. All this for the cost of your discarded coffee grounds and vegetable scraps.

Bees make honey, which gets a lot of money at market. Mealworms can be used to feed other animals and are valuable to pet supply stores. Edible snails have a lot of potential for providing income via restaurants and gourmet stores. For the particularly brave homesteaders, mealworms, snails, and crickets are all edible, and, by many accounts, quite tasty. So it's possible to raise one of the most concentrated protein sources in the world in a space the size of a closet.

Miscellaneous

Finally, don't forget about the unusual and often ignored animal products. Chicken, duck and goose feathers are sold to clothing and bedding suppliers.

There's also a market for peacock and emu feathers. Rabbit skins, if processed and dried properly, can sell for a few dollars each. Products like honeycomb can be used by local craftsmen and provide an alternative revenue stream for beekeepers who may be interested in making and selling candles and/or lotions.

While most of these ideas aren't enough to justify raising a particular animal in and of itself, a little extra income can often be the difference between breaking even or losing money. Even better, a little extra income can often be the difference between breaking even or making a profit. A profitable homestead is a lot more sustainable than an unprofitable one.

Keeping your animals healthy

There are tons of resources on how to care for the health of your animals (I'll list a few of them shortly), but honestly, about 80 percent of animal health boils down to cleanliness, ventilation, fresh water, and plenty of food.

Through thousands of years of domestication, most breeds of animals are hardy and relatively easy to care for. If you can keep their enclosures clean and give them enough ventilation to allow for proper airflow without drafts, then disease and discomfort will be kept to a minimum. Water containers should be rinsed daily and cleaned at least weekly.

Animal Shelter Basics

While there are some specifics to different animal species and their shelter needs (rabbits do a lot better with tunnel space and resting boards, for example), there are some basic requirements all animal enclosures must have. Some are obvious, like a roof, while others not so much. Here are the basics you need to know.

Ventilation—When it doubt, ventilate. Air flow is important for animal enclosures. Because animals don't usually leave their enclosures to defecate, the mold and mildew that can build up in the feces inside the enclosure can cause respiratory ailments. Proper ventilation helps prevent this by preventing fungal growth.

Room to move—Without the ability to exercise, your animals will get sick and possibly do harm to themselves or others. Some animals don't need much space, but others, like chickens, require a run large enough for them to exercise, socialize, and more.

Security—Enclosures have two jobs: keeping animals inside and keeping animals out. Your goats won't live long if predators can get inside their enclosure. This doesn't mean your space has to be Fort Knox. Most predator attacks happen at night, while humans aren't around to intervene. The easiest way to deal with this is to lock up your fur and feather babies in the evenings. It'll keep them away from harm and save you money on costly fencing. That doesn't mean you can skimp, however. I don't know what chicken wire is for, because it's almost useless for keeping chickens safe.

The water should always be replaced daily. For everything else, check with your local veterinarian.

Most large animal vets can provide all the resources on how to care for cows, alpacas, and the like, but for the following animals, sometimes it's difficult to find a veterinarian experienced in their care. This isn't the fault of the vet. Many of these animals just aren't cared for in the same way as a cow or horse might be; chickens, for example. Most farmers simply cull a sick bird and bury it in the compost pile. Economics mean some animals just don't receive an equal level of care. But, ethically, I believe we should do as much as possible to care for the animals in our stewardship, so here's my favorite list of animal care resources, specific to those "neglected" species.

Chickens—A blogger, Lisa Steele, produced a book in the last few years named after her blog. *Fresh Eggs Daily* is a concise little volume that outlines the care and health needs of chickens from a natural and holistic viewpoint. It's a short and invaluable read.

Goats—While there are some excellent books on the market about goats and goat care, the best resource for goats is Fias Co Farm (faiascofarm.com). The site is a wealth of information, with just about every question on goat health and care asked and answered. While Molly, the owner of the site and farm, produces an herbal wormer, the site provides both natural, holistic remedies as well as conventional remedies and treatments.

Rabbits—The aforementioned book by Chris McLaughlin, *Hobby Farms: Rabbits: Small-Scale Rabbit Keeping*, is still my favorite resource for all things rabbit. It provides tons of information about the care and health of these little animals.

And, finally, don't discount a vet for the above. Call around and ask neighbors and other homesteaders. While vets for the above animals aren't always knowledgeable, many vets are, in fact, aware of some of the less popular livestock animals and are more than willing to work on them. For some animals, such as rabbits, vets are used to caring for the pet varieties, which gives them the experience necessary to get you the help you need.

Keeping your animals happy

Finally, one aspect of animal care that I feel is often overlooked is animal happiness. A lot of the time, especially when raising animals for meat, it's easy to depersonalize or minimize an animal's emotional well-being. Ethically, I believe that as homesteaders and farmers, we have an obligation as stewards of the animals we partner with to provide them with comfortable living arrangements, quality food and care, and, when necessary, the proper stimulation to ensure as low stress and happy an existence as possible.

Outside of the ethical considerations, there are pragmatic concerns that indicate we should be a bit more concerned with the happiness of animals.

Research has shown that stressed and frightened animals produce less milk and fewer eggs. The meat of frightened and stressed animals is of lesser quality and doesn't taste as good. And it's not difficult or expensive to provide some sort of care that allows for animals to live happier lives. So, please, take some time to find out what makes your animals happy and try and

Don't overlook happiness as part of your animal's overall health. Take time to find out what makes them happy and try to provide it for them.

provide it for them. That level of care will not only pay off in increased yields and higher quality, but it will let you exist with them in such a way that is morally, ethically, and, ultimately, spiritually rewarding.

Read more about it

For more information on the ethics of providing happiness for animals, read any of the books on animal welfare by Dr. Temple Grandin. Grandin is a pioneer of animal welfare who has done invaluable research on the well-being of animals, focusing particularly on livestock raised for slaughter.

RECEIPTS

CHAPTER 8: RECORD KEEPING AND PLANNING

Don't skip the planning stage

When we first started homesteading, the first thing we did was overanalyze every decision we made. Then, partly in frustration, we just started doing stuff. We took our little bit of knowledge and started applying it to everyday situations.

We had chickens, and we knew that chickens needed four ounces of food a day. So we made sure the chickens had four ounces of food a day. I weighed it out and figured out how much chicken feed a coffee can holds, and then made sure the chickens got it.

Every day, four ounces per bird without fail. We were very, very proud of ourselves. Since our chickens were well fed, they started producing more eggs. And we liked eggs. So we got more chickens.

Getting chickens at the time was easy. My sister-in-law sold the eggs we produced at the retail store where she worked. Customers and coworkers got to know her as the chicken lady. So, whenever anyone got chickens, they'd talk to her about them. Quite a few times, because of moving, or because they'd bitten off more than they could chew, people would give us their chickens. Which we, of course, accepted.

At the end of one summer, we had one hundred and twenty birds in our flock, and the feed bill was about $300 a month.

It was officially costing us extra to homestead. That wasn't the kind of homesteading we were looking for. We wanted a homestead, not a hobby farm.

So, we sat down and made a plan. We set up a fodder system to feed the birds we had, started culling the flock and filling our freezer, and expanded the enclosures and coops. Within a couple of months we had cut our feed costs to about thirty dollars a month and actually started producing more eggs. Because the birds weren't as stressed, were fed better, and had more space to move around in, they produced more.

We had meat in the freezer, eggs in the fridge, and a better grasp on how to handle things.

The entire ordeal could have been avoided, however, if we hadn't skipped the planning stage before taking in everyone's chickens.

Planning is probably the biggest fail made during the beginning of a homestead. While it's expected—and even encouraged—that you'll make mistakes at the beginning of your homestead journey, proper planning is often the difference between quitting and continuing.

Conversely, you should know that your plan isn't set in stone. A good plan is flexible and above all serves as a starting point. A good plan—as illustrated

Don't skip the planning stage. Getting chickens was easy when we first started homesteading, soon we had more than one hundred birds and a three hundred-dollar feed bill.

above—will allow you to experiment, and fail, without going broke first, which will ultimately give you the freedom you need to achieve success.

Making a farm plan

Since you're dealing with a homestead, you'll actually need two plans: a farm plan and a business plan. We'll discuss the basics of business planning in a later chapter, but for now you'll need to start with a farm plan. (Note: Some people combine their farm and business plans. I don't, as homesteading tends to be a mixed-income enterprise, with income coming from crops, handicrafts, and even services like classes, I prefer to do a farm plan separate and later incorporate that into an overall business plan.)

Your farm plan should include two parts: animals and plants.

The plants are the crops you'll grow, including herbs, flowers, and vegetables. The animals are the animals you'll raise for profit

Be sure to include a calendar for harvesting in your farm plan.

or food production, with doesn't include pets, be they dogs, cats, chickens, or goats.

First, establish your goals. How much money do you want to make? Or, if you're farming for calories (more on that in a moment), how many calories would you like your farm to produce?

Start modestly; if you under promise and over deliver, even to yourself, you're much more likely to stick with it. It's a lot easier to grow over time and succeed than it is to overextend and then try to figure out how to pull back.

To start with, your farm plan should begin with why you're engaging in a farm operation to begin with: cash or calories. If you're farming for cash, or profit, you'll need a slightly different plan that if you're farming for calories. In other words, farming to feed you and your family.

Starting with crops, your farm plan should include a list of crops you'd like to grow, a copy of your soil test, a list of amendments you'll need for the soil, a calendar noting when to add those amendments, a calendar for planting, a calendar for harvesting, and a watering schedule. Your farm plan should also include a budget so you don't overspend before you get anything in the ground; and, yes, this includes farm plans for calories.

You'll also need a plan for your harvest, and this is the big split between farming for calories versus farming for cash. If you're farming for calories, you'll need to decide what you will do with your harvest. Will you freeze your crops, ferment them, or can/preserve them? If you don't know ahead of time, you'll find yourself in the situation I was in—I had three hundred pounds of squash on my kitchen table with no idea of what I was going to do with it. If you're farming for cash, you'll need a plan to get those crops to your customers, be it CSA pickups, deliveries to restaurants, or direct sales at a farmers' market.

Planning for your animals is similar. You'll need a list of the animals you're going to keep, a plan for their shelter and food, as well as a budget for how much the animals will cost. Again, you'll need a plan in place to process the products of those animals. If you're raising them for meat, you'll need some sort of space and method of doing so, or at least some idea of how much it will cost to have them slaughtered and butchered at an off-site facility (Note: Some states allow for the home processing of birds, but nearly all of them require a USDA-certified butcher for processing most other animals. Check with your local extension office to find out more).

You'll need to know how to get the meat—or the eggs, milk, and honey—to market if you're raising them for sale. If you're raising them for calories, you'll need to figure out how to store the products, either in your freezer or some other

Regulations

It must be noted that most animal products—eggs, milk, meat, honey, etc.— have strict regulations on both the state and federal levels. For example, most markets require eggs in America to be washed and kept at a specific temperature to avoid spoilage. These regulations are in place to protect consumer safety. You can contact your local farmers' markets to find out more about these regulations, your local health department, or your local cooperative extension office.

Whether you use a simple ring binder or keep everything on a computer, when it comes to creating and keeping your farm plan, choose the method that works best for you.

When designing your farm, include the fields or beds you're going to plant and the rough measurement for each one. Smaller beds provide multiple benefits.

method. Meat can be canned and so can milk. Eggs can be frozen. Either way, you'll need a plan to deal with the products when they're ready. You don't want to be in a situation where you raise broiler chickens and then don't know how to process one.

I find it best to keep all this information in a three-ring binder. For me, it means I can carry my farm plan anywhere without having to worry about an Internet connection or power source. So, whenever I go to the extension office to get advice or input, it's right there. And since I'm an organization nerd, I can outfit it with indexed tabs, a table of contents, and even stickers!

There is a caveat to my three-ring binder system: I keep all that information backed up on my computer. So, my spreadsheet for my crop list, my calendar, my soil test, all that information is on my laptop. That way if I have to make any changes, especially early on in developing my farm plan, I take notes directly on my hard copy, and then make the changes on my computer when necessary. If I didn't do that, my plans would be a squiggle-covered mess well before I started to implement them.

Farming versus gardening

When homesteading, the difference between farming and gardening gets blurred. Chances are you won't be working with the dozens to thousands of acres a traditional farmer works with, but that doesn't mean you aren't farming. For me, the difference between the two is the intention of your outcome.

In my experience, gardeners tend to be driven by experimentation. It's not so much about growing food for your family or making money, but about trying to find out exactly what a garden can grow, with a point of pride for the more obscure and difficult-to-produce plants.

This doesn't make farming superior to gardening. In fact, it could be argued that gardening is a purer expression, as opposed to the cold application of farming. The best thing to do is not worry about labels. If you're homesteading, you'll be adapting methods from both farms and gardens and using them for whatever you need. Again, there's not a litmus test on what homesteading means.

Farm/garden design and layout

One of the things you'll have to include in your plan is a layout of your farm. The layout should include the buildings on your property, water sources, the locations of animal enclosures, fields/beds you're going to plant, the rough measurements of each field/bed, and some sort of identification for your beds, like letters or numbers. This way you can identify the locations where you'll plant specific crops later in the plan.

While designing your layout, I suggest you use a zone design method from the permaculture movement. This method puts whatever will take the most work to maintain in the spaces closest to your home. Anything that will be harvested quickly, animals that require extra care, etc., will be located closest to your front door. The items that require the least maintenance, things like fish ponds or fruit trees, will be farther away. The idea is that putting high-maintenance items near you allows you to expend more energy working on them as opposed to traveling to them.

Also, and this this is a personal lesson, it's better to make more small spaces than it is to make one large space. For example, if you have the space for a one hundred fifty-foot-long row, you'll want to make six twenty-five-foot-long rows.

Psychologically speaking, making a shorter row or smaller bed is easier to work on and around. Also, by breaking your growing and planting spaces into smaller areas you can increase biodiversity while creating boundaries between similar crops, slowing the spread of disease.

When you get your farm plan together, don't stop. Your farm plan should be an ever-growing body of information and work. So, as you follow your plan, you'll take notes and keep information about costs and how the plan changes. You also want to keep weather information so you can compare the weather patterns in your area over time. All this information will be used in the following year's farm plan.

Here's a list of data and information to keep with your farm plan.

- Receipts—Every cost incurred on your farm should be tracked. Keep the receipts for everything you purchase so you can properly budget next year. In addition, track the information in a spreadsheet or ledger so you can use it for your taxes, if necessary, at the end of the year.
- Crop yields—Separate this by variety as well as crop. Also keep track of the yields in specific fields and beds so you can better plan where to plant next year.
- Weather data—Tracking the temperature and rainfall on your homestead will allow you to create a clear picture of your specific microclimate, so you can use it to your advantage in the coming years.
- Personal moods—While it may seem trite, by tracking your energy levels

and morale during the year, you can plan your work schedule to better suit you in the coming years. For example, if you find yourself frustrated, exhausted, and hot in July, plan for crops and animals that don't require as much care during that month. Or, look for ways to reduce your workload, mulching beds to cut down on weeding, for example. By doing this you can better care for yourself, which lets you better care for your homestead.

- Income—Track all the money you make from your homestead and where it comes from. This will give you a better idea of where to place your energy and efforts next year. If your eggs produce 15 percent of your homestead income, and your vegetables only produce 10 percent, it might be more feasible to put more energy into egg production.
- Calories—This one is a bit tricky, but if you're homesteading specifically for food, you'll need to figure out how many calories of a specific product is produced. Again, just like income, this will let you know how to expend your energies more efficiently. Lettuce won't sustain life as much as, say, potatoes, so you might expand your potato patch at the expense of your leafy greens.

You can do both, but you should determine if your main priority is growing food to feed your family or for income.

What to grow: cash versus calories

Finally, while you don't have to choose between homesteading for calories over income, or vice versa, what you decide to focus on will be a mitigating factor in what you plant. A friend of mine spent the first year of vegetable production providing food for about twenty CSA boxes. For weeks she harvested vegetables and drove them out to her CSA drop-off point. She got paid for the vegetables, and then immediately drove to a nearby grocery store and spent the money on organic produce and food. After a few weeks of doing this, it finally dawned on her that she was literally growing the food to sell, only to turn around and buy the exact same food at a higher cost.

Since then, she's decided to focus on growing more food to eat until she can expand her operation to include food shares for customers off the farm.

Here's a breakdown of particular crops and whether they're better suited to grow for calories or cash.

Cash crops

- Cut flowers—These usually can't be eaten, and even the edible varieties don't provide much in the way of calories. The markup, as well as the fact that small, organic farmers tend to grow more food than flowers, means this crop can fill a niche in your community while providing you a paycheck.
- Herbs—Herbs provide lots of flavor, can be grown in small spaces, and sold at a nice rate at market. Again, they don't provide much in the way of calories, but be aware, they add flavor to any meal and often have medicinal value. So, if you do sell herbs at market, bear in mind you don't want to sell your homegrown "medicine cabinet" in the process.
- Vegetables—The most profitable vegetables to sell at market tend to be the lowest in caloric value. Think salad greens, spinach, and kale. These crops, while sometimes nutritionally dense, often do very well at market.
- Meat—Organic meat draws a good price. I've seen grass-fed beef sell for eight dollars a pound. While meat produces a lot of calories and tends to have greater regulations attached to it, the price you can fetch for meat means that, if possible, it's worth having a portion of your meat production for market sales.

Calorie crops

- Potatoes—All potato varieties are calorie dense, store well, and grow in nearly every kind of soil. While "new" potatoes do well at market, by leaving them in the ground longer, you'll produce a lot more calories over a season.

A note on insurance

For your protection, if you're going to sell food from your homestead, invest in liability insurance. Liability insurance, which can be obtained surprisingly inexpensively, provides you a bulwark against lawsuits that could occur in the case of food poisoning or other issues. Not only does it protect you, but it signals to members of your community that you're concerned about doing business in an ethical and professional manner with an emphasis on safety. Many markets, including grocery stores that purchase and sell locally produced food, require liability insurance, meaning that spending a few hundred dollars a year could lead to much more profit for you and your enterprise.

- Eggs—Eggs are often considered a cash crop, but since many egg producers don't charge enough, it's probably better to consider eggs a calorie provider. A good sized egg can provide ninety calories, and chickens, if properly managed, aren't too expensive to feed.
- Honey—Honey gets a really great price at farmers' markets, but since honey is literally made of sugars, it packs an awful lot of calories. You might consider putting your first couple of years of honey up and only selling the

beeswax and royal jelly, if possible. Then, after you have about five to ten gallons stored away for personal use, start selling it.

- Dairy—Milk provides a lot of calories. A good milk goat can produce a gallon of milk a day, which is approximately 2,668 calories of fat, carbohydrates, and proteins. Milk is also arguably more regulated than medical marijuana, which means that by the time you set up a fully operational dairy, complete with a pasteurizer—which most states require before sales are allowed—the cost ensures you won't see a profit for years, if ever.

My Favorite Crops

Every year I do an overly ambitious farm plan. And every year I adjust and add to it in hopes of finding a magic crop that I can put in the ground, never water, never weed, and produce a bumper crop. So far, I haven't found it. However, I have found some really amazing crops that I grow every year. Here they are.

Popcorn—My favorite, thus far, is a miniature blue popcorn. The interesting thing about popcorn is it doesn't require the pesticides sweet corn requires. Because you aren't interested in producing whole ears, you can break off the bad parts of the ear and still get a good crop. Popcorn's also beautiful: All the decorative varieties you see around Thanksgiving and Halloween are popcorn. Varieties include Glass Gem Corn, "Indian" Corn, and Sacred Apache Corn. In addition, if you grow it as a trellising support for beans, cucumbers, or even small winter squashes, you can maximize space to the point where you don't even need to sell it. It's also a dried product, so it keeps for years. And every variety of popcorn I've seen can be ground into meal for cornmeal or grits.

Jerusalem artichokes—Resembling a cross between echinacea, daisies, and sunflowers, these tall plants produce a starchy tuber that's actually pretty tasty. It's beautiful and grows well in just about every type of soil. Almost too well; it's considered invasive in many areas. If you plant it in a bed or in the ground, you'll never get rid of it. However, if you grow it in containers, you'll have a pile of delicious tubers before you know it.

Walking onions—Also known as Egyptian walking onions, these are another "invasive" species, but not nearly as invasive as Jerusalem artichokes. Walking onions, unlike most species of onion, actually produce bulbs at the end of their stalks instead of flowers. When the bulbs get heavy enough, the stalk falls over and plants itself in the ground. The onions are small, but they can be used like green onions or shallots. Plus, they plant themselves; how great is that?

Blue potatoes—I like planting potatoes in general, but once I discovered I could plant blue potatoes, those are the only ones I grow now. They're potatoes, which are pretty low maintenance, but they're blue. That's it. Having blue starches that taste incredible in the ground just makes me happy.

3 | Section Three: Dollars and Sense

FARMER'S
SHOP

CHAPTER 9: A BUSINESS PLAN

What is it for, and do you need it?

Honestly, for me, business plans are no fun. I'm not a numbers person, and the idea of a profit and loss statement makes me break out in hives.

But, I cannot argue with a business plan's necessity in a homestead. Even if you aren't homesteading with an eye toward earning money, you should consider creating a business plan anyway. A business plan lays out, in plain numbers, how much you're spending on your homestead, and how much your homestead brings in. You can adjust that model to reflect how many calories produced, or how much money you save on groceries and so much more, using both those numbers as a replacement for "income."

You can also consider other things as income: products or services received as a result of bartering, money saved by producing your own soap, detergent, medicinal herbs, or handmade gifts for friends and family members.

If you're homesteading with an eye toward producing a profit, a business plan will be invaluable when seeking financing, loans, grants, even crowdfunding campaigns. Armed with a business plan, you'll be able to show anyone who asks that you're serious about producing income on your homestead and exactly how you plan to do it. It will also give experts and friends ways to contribute. For example, if they see that you're planning on starting a cut flower operation, they can suggest market niches and flowers that you might not have considered. Outside insight is invaluable in any business enterprise, and a business plan lets you get that insight by providing others with a clear view of exactly what you want to do.

Your business plan should include the following:

- Your background, including the background of your partners. Include everyone who will be involved in your homestead in this list. Is your twelve-year-old daughter an artist? She'll make the signs. Does your husband have a background in construction? He'll build the sheds and fences. In whatever manner your family and partners will contribute, make sure these items are included.
- A budget, including projected costs. This is probably the most time-consuming aspect of business planning, but by outlining costs at the beginning of your year, you can plan ahead. You can also make a determination of how much money you'll need. With that information, you can judge how much of your savings you'll need to invest. You can also make determinations for outside funding sources: the aforementioned loans, grants, and crowdfunding campaigns.

- Your concept, or, exactly what your business will be. For most of you, this will be a homestead, and, as you'll see below, homestead businesses have a lot to offer communities.
- Your method for implementing your concept, or what you're going to do. Are you going to grow crops or buy and resell crops (and that's a possibility, especially if you decide it would be a better service to your community to provide a co-op for other farmers as opposed to starting a farm yourself)? Are you going to make lotions and soaps? Are you going to crochet blankets or sew baby bibs? What about some combination of the above? Homesteading businesses are good at providing multiple items. Think of your homestead in these holistic terms and you'll have a better chance of finding a niche in your community.
- Your goods and services. Make a list of everything you'd like to provide your community under your concept, and don't leave out services. As a homesteader, you can host gardening and how-to classes at community centers or at your house. You can teach crochet at the local senior center. Some homesteaders are yoga instructors on the side. You can also contract with area homeschool groups to provide tutoring or classroom activities. Be creative and confident: you'd be surprised at what a dedicated homesteader can do that most members of the community are fascinated by or completely ignorant of.
- Where your goods and services will be sold. Where you'll be setting up shop and how you market your goods and services are integral parts of your business plan. You can be the best and fastest baker of homemade bread in a tri-county area, but if you don't have a way of reaching your customers, you won't have a business. While the obvious one is the local farmers' market, don't discount other methods. Check out smartphone apps, like Farmzie, which is a virtual farmers' market that allows customers to order direct from farmers. Look into food hubs in your area. Walk into a grocery store and ask if they'd be willing to buy local food. Check with health food and co-op stores to see if they have space for homemade soap. Check with community art programs for craft fairs and museum showings; a piece of folk art, like a handmade quilt, sold for the right price can make a homestead profitable for an entire quarter. Talk to homeschool groups. Talk to other homesteaders. Reach out to people. Start an email newsletter. Talk to anyone that might be interested in what you do. Do it with confidence, because homesteaders are a pretty amazing lot, so I promise you'll find someone interested in investing in what you do.

While you can put together a business plan without using third-party resources, you might want to consider contacting an outside party for help. See if you know entrepreneurs, accountants, or bankers in your list of friends and family. They will be able to help you put together your business plan in a manner that's attractive to potential investors.

If you don't know anyone who can provide you with guidance on the creation of a business plan, the University of Minnesota offers AgPlan, a free online app that allows users to create business plans. Although it was designed specifically for agricultural enterprises, it can be used for any type of business. Sample business plans and tips for what to include in a business plan are provided. Once completed, users can also download and print their business plans.

While at first you might not be interested in turning a profit with your homestead, it's still a good idea to start with a profit-driven mindset. If you view your homestead operation with the cold, draconian logic of a business, you can pursue your passion and still provide for the basic needs of your family. Without some sort of income, or large savings, you won't be able to spend time homesteading. Instead, you might find yourself in a situation where you have to take an outside job. There's no shame in that, and if you find your business endeavors dropping off, it's not a failure to pursue income by other means. Don't go bankrupt to live your dream; it's unnecessary on a lot of different levels. Also, working off the homestead provides a stable income that doesn't often come from working for yourself. There's something to be said for the security of a weekly paycheck.

Determining profit . . . and its importance

How do you determine a profit?

When approaching a business, some people attempt to set a profit margin: a percentage of profit based on what you earn after paying expenses. A profit margin of 20–30 percent is usually a good point to hit. For example, if you determine your homestead operation will cost $20,000 a year in expenses (including property

Even if you're not concerned with making a profit, it's still a good idea to assign a dollar value to your efforts.

purchase or rental), then you'll want to hit $22,000 a year in total sales.

For homesteaders, however, these numbers don't always work out. When determining a "profit" for my homestead, I start with expenses, add 20 percent, and work from that point. So, if I determine I need $1,200 each month to pay my bills, I add $240 to that number, and then determine how I'm going to achieve that income. So, in this case, I'd start with a goal of $1,440 per month and do everything legally possible to make that money via my homestead. Although it may seem like a very similar approach, I find that taking it from a backwards view, it's often easier to digest, less abstract, and not as psychologically daunting.

Conversely, if you're in a situation where you're not concerned with making a profit, it's still a good idea to assign a dollar value to your efforts. For example, if you decide you want to grow 40 percent of your food, a good way to approach the profit issue is to determine what 40 percent of your food costs. So, you'd calculate your average food bill over a month, let's say $500 for ease of calculation. That means 40 percent of your food bill would be $200 per month. So, to achieve a "profit," you'd have to spend less than $200 on producing your own food per month.

Doing this prevents you from winding up in a hobby farm situation, where, ultimately, homesteading becomes a hobby instead of a lifestyle. While there's nothing wrong with that, knowing that ahead of time allows you to approach it clearly from the beginning, making it more sustainable over a long period of time.

Don't forget about bartering. Bartering is one of my favorite ways to do business, especially with fellow homesteaders. I've traded eggs for vegetables and meat. I've traded work for goods. I've traded seeds, and so much more. My favorite was the arrangement where I bartered duck eggs for home-brewed beer.

There's always something your homestead produces that can be bartered. I've got wild blackberries growing on my homestead. I can promise you that once I get a solid harvest, I'll be bartering homemade wine for all sorts of things.

Bartering is fun, it's something we can all do, and it creates a truly local economy. Bartering can be a part of making a profit as well (Note: The IRS does consider bartered goods to be income.)

When you barter with an eye toward making a profit, you will want to establish a value on the goods you are getting and a value on the goods you're trading. In the case of bartering, however, the value of the goods is about as close as you can get to the "real" value of a thing. To me, my blackberry wine is worth about eight dollars a bottle. I will never sell it for eight dollars, but that figure is what I would spend producing each and every bottle of wine. To make a "profit" bartering, I would have to barter for something that I perceived to be worth more than eight dollars for each bottle I produce.

That doesn't mean I get one over on the individual I'm engaging in barter with. The goal is to provide both parties with a mutually beneficial experience. In the case of my wine, a person would consider how much she would pay for locally produced blackberry wine, which might be ten dollars, so she'd want to give me approximately ten dollars' worth of, say, fleece for the wine. That fleece wouldn't cost her ten dollars to produce, but we both walk away with ten dollars' worth of material, so we both profit from that exchange.

With a little imagination, and a lot of work, there's no reason why you can't make your homestead financially sustainable, and that's really the purpose behind creating and implementing a business plan.

Should profit be a motivator?

There's a company called Tumbleweed Tiny House Company. The company builds and sells tiny houses (about 200 square feet in interior area) and plans so individuals can build the tiny homes themselves. *From Scratch* magazine contacted the company and agreed to cover one of their workshops where they discuss every aspect of how to build the homes. The workshop was being taught by one of the company's instructors, Art Cormier. During the workshop, one of the students asked Cormier about the company's ability to make a profit by selling the plans instead of the homes. Cormier assured the student that Tumbleweed was, in fact, making a profit, and then he said something that has been a guiding principle for me personally and for *From Scratch* magazine.

"You know, profit is not a bad thing," Cormier said, "but it's a terrible motivation."

So, while you're creating your business plan and thinking about your homestead and what it means, bear in mind that "profit" is not a dirty word. But keep Cormier's words in mind; it is a terrible motivation, and you'll have to find something more concrete, if you will, to serve as a purpose for your homestead, be it feeding yourself and your family, serving as a steward for your land and animals, or providing your community with fresh, local food alternatives. Whatever it is, make sure you can live with it, and that you can use your profits as the tool they are designed to be.

CHAPTER 10: FINANCING

The basics

One of the biggest issues you'll run into regarding homesteading is funding it.

Even if you have the property and all the materials you need to build, there will still be a period when you have to provide for yourself and your family while waiting for the operation to at least pay for itself in whatever way you consider valuable. For that, you'll need capital. Honestly, I didn't know that when I started homesteading. I was under the impression we could just find our little piece of land, get some goats, put seed in the ground, and just go off-grid.

My intent, silly as it sounds right now, was to find a way to survive and raise my family without money.

It doesn't really work that way.

My family needed money to live, and we needed money to get our business started.

Funding to start *From Scratch* came from our savings, part-time jobs, and culling our lives down to the bare essentials.

Funding for our homestead in North Carolina came from a similar situation. Lots of saving and working and scrimping.

Both methods work, but there are other ways we didn't really know about until late into our homesteading journey. While we don't regret the circuitous route it took for us to finally get our act together, it would have happened a lot faster with some new sources and some forethought.

Traditional funding sources

First, let's discuss traditional funding sources: loans and grants. A loan is money given to you with the expectation that you'll pay it back with interest over a set period of time. Grants are funds given to you if you or your operation meets certain criteria. The money provided via grants does not have to be paid back.

Loans

Loans are easier to get, for the most part. Traditional loans are from banks and usually come with a fair interest rate. Most farm loans—homestead loans are going to be considered farm loans—are going to be tied to property. Which means you'll be securing the type of loan known as a mortgage. Mortgages are usually paid out over thirty years. Interest rates for mortgages, however, are usually relatively low compared to other types of loans.

Most homestead loans will be considered farm loans and will be tied to the property.

The United States Department of Agriculture and the Farm Service Agency also issue agricultural loans. In addition, some states have agricultural programs that also issue loans.

However, these loans aren't usually a good fit for homestead operations. To be eligible for an FSA loan an agricultural enterprise has to provide tax returns showing a profit for at least three years before applying for the loan.

State agencies sometimes have seemingly confusing rules outlining what can be considered a farm or agricultural enterprise, often including minimal land requirements. This means that a lot of homesteads don't qualify for the loans and grants from state agencies right off the bat.

Grants

Grants tend to require a bit of legwork to get. Generally provided by government agencies, grants are usually offered for very specific needs. It's difficult to get a grant as a start-up operation. After two or three years of getting your homestead going, you may be eligible for all sorts of money. Grants, however, are highly competitive and often take a lot of effort to put together. Each grant project must include a proposal on how you'll use the grant money and details about your operation. It's a very time-consuming process, which is why many applicants for grants work with a professional or semi-professional grant writer.

Grants are the primary way the USDA and other government agencies encourage agriculture in America. Government money is available for all sorts of reasons, but if you get it, you'd better have all your ducks in a row.

Loans and grants are both great sources for developing capital, but they have their drawbacks. The idea of a thirty-year mortgage scares the dickens out of me. The only commitment I want to last that long is my marriage. Grants are fine, but in my experience by the time you realize they're available, the deadline has come and gone.

So, what other options are there?

Nontraditional funding sources

Crowdfunding

Crowdfunding is a newish way to raise money outside of traditional loans and grants. Essentially, you take to the Internet, providing a business plan and model that investors can choose to contribute to. The investors tend to provide small amounts, some as small as a dollar, that can be used to fund your concept. Most crowdfunding sites require users to establish a funding goal. Be aware: if that goal isn't reached, you often don't get any money. So if you have a goal of $2,000 and that goal isn't achieved in the time allowed to raise it, you get nothing, no matter how close you are to reaching the goal. However, if you surpass your goal, you do get that money.

Most crowdfunding sites suggest that you offer your investors "rewards" for their investments. For example, if you're starting a rabbit fleece needle felting business, you could offer dryer balls for anyone donating more than, say, twenty dollars. While the people who contribute to crowdfunding campaigns are called investors, they don't actually get a return on their investment. Crowdfunding investors usually invest money into a project out of a sheer desire to see the business get off the ground. The most popular crowdfunding services are kickstarter.com and gofundme.com, but since you're a homesteader, you might want to consider barnraiser.us, a crowdfunding site devoted to agricultural and food enterprises.

Microloans

Popularized by Nobel laureate Professor Muhammad Yunus, microloans were originally conceived in the eighteenth century, but in recent history became a way to alleviate poverty in Third World countries. By providing small loans—loans too small to be profitable for traditional banks to—to individuals and businesses, poverty in these countries could be alleviated by allowing individuals to create small businesses so they could earn a living for themselves and their families. Since then, the concept has taken off in the United States.

The U.S. Small Business Association and the Farm Service Administration, as well as private, nonprofit groups, have begun providing microloans to people all over the country. The loans are typically less than $100,000 and sometimes they're less than $1,000. The interest rates are often lower than bank loans, and they may have a shorter payback period.

Microloans are better suited to established businesses and operations, but they allow an expansion of opportunities for many small businesses. I know of a farmer in North Carolina who used a $700 microloan to purchase a brooder, allowing him to start a small chicken hatchery. Organizations like Slow Money (slowmoney.org) actually allow private individuals to provide microloans to borrowers at low interest rates.

CSAs as a funding source

The most unusual, and fun, funding source to me, however, has to be the CSA model. CSAs, or community supported agriculture, are like local crowdfunding campaigns.

At the beginning of the year, farmers sell shares of the harvest to community members. That money is then used to provide for everything the farm needs to get that harvest in the ground.

A CSA is a nontraditional way to fund your homestead, and it provides the added benefit of building community.

Then, once the crops and harvests are ready, CSA investors pick up their share of the produce.

This way farmers are able to finance at least part of their operation at the beginning of the year, which tends to be the time of year when capital is lowest in agricultural enterprises, while consumers get a voice in what the farmer produces. CSA shares tend to be, but aren't always, cheaper than buying the produce piecemeal. The drawback for the consumer, however, is that you don't have a say in what you get in your produce box, for the most part. When a particular crop comes to fruition, that's what you get. The great thing about that, though, is you're always eating produce in season, which means they're going to be the freshest, tastiest veggies ever.

If you're interested in doing a CSA, start with your business and farm plans. Find out how much money you'll need to get a crop in the ground, and then sell shares based on that. Most CSA shares sell for about $300, but that number varies based on your region. So, for example, if you need $4,000 to produce a crop, and CSA shares sell for about $200 in your region, you would need to sell twenty shares to hit that goal. CSAs usually provide shareholders with produce shares for a set number of weeks during the season (usually at least eight). While the CSA shareholders expect their shares during those weeks, it doesn't mean farmers can't sell produce during other weeks.

In addition, many CSA farmers overproduce, which means they can provide shareholders with their produce, and then sell overflow produce to wholesalers, at the farmers' market, and to restaurants.

In addition to vegetables, do not neglect other farm products for your CSA box. You can include eggs, honey, meat, bread, cut flowers, herbs, even handcrafted items like soap. If you do, make sure your CSA share price reflects these extras. While CSA shares are considered direct sales, they still fall under the same rules and regulations of other farm products in your state.

The best thing about a CSA, however, despite the monetary benefits, are the interactions you'll have with your customers. You get to know people in your community and find out what they like.

You learn a lot about a person when you're feeding them eight or more weeks out of the year, and the friendships you make will be invaluable as you continue your operation.

Don't forget to keep the community in community supported agriculture. Whenever you can, try and do something nice for your shareholders: Treat them to extras, host a bonfire, or invite them out to your homestead.

Every chance you get to make a connection with your shareholders—all of your customers, really—is a chance to make new friends and expand your business.

Be aware, however, that if you're starting a CSA for the first time, a bad failure can ruin your reputation before you even have the chance to get started. So, start small and build slowly. As I said before, it's better to under promise and over deliver.

SWEET CORN
TOMATOES
WATERMELONS
PEACHES
FARM MARKET
NEXT RIGHT
MON-SAT 9 TO 6

CHAPTER 11: MARKETING

Getting the word out

So, you've got a farming method, a farm plan, a business plan, and funding sources. The only thing left to do is tell the world about it through marketing and advertising. Marketing, in its simplest and purest form, is the act of letting people know what you do and how it has value.

Honestly, for most homesteaders, this isn't rocket science. All you need is a good dose of extroversion, a name for your homestead, and a business card.

Homesteading, in many ways, is romanticized by the general public. The type of agriculture practiced by homesteaders, along with the rugged individualism that tends to be associated with it, has been mythologized since the beginnings of the United States. President Thomas Jefferson pointed out that yeoman farmers (small homesteaders) were the crux of the American Revolution.

Laura Ingalls Wilder waxed poetic about her homesteading family in eight books, not counting the spin-offs, children's books, chapter books, and more that have sold well since 1932 (and that's not counting the television show). Going back even further, the "back-to-the-land" movements actually have their roots in the Roman empire. My favorite homesteader from history is Lucius Quinctius Cincinnatus, a Roman dictator who was literally called from plowing his fields on his small farm to lead the empire during a time of crisis. He returned to his little homestead a mere fifteen days later, having saved the country from destruction).

Since all that groundwork has been laid, all a homesteader has to do is explain to others what she's up to, and everyone is on board.

But, what if you aren't an extrovert?

To put it bluntly, you have to work on that.

Until then, you can recruit other members of your family, friends, or even hire someone if you can afford it. Whichever you decide to do, make sure whoever has the responsibility of serving as spokesperson for your operation is front and center whenever your homestead business has to interact with the public.

In the meantime, get a logo designed, business cards printed, and have some signs made. It doesn't have to be polished or even professionally done. If you can get one of your kids to paint the name of your homestead on a board and that's all you want to do, then that's perfectly all right. You just need to make sure that the people who want to know more about you and your homestead can find out. Later, as you get more comfortable with tooting your own horn, you can start to produce other marketing materials: flyers, direct mailers, recipe cards, whatever tickles your fancy.

If you make enough money, you can start to place ads in the local papers, radio stations, or even on television. Or, conversely, you don't have to do much more than the basics: business cards, homestead name, and signs.

If you can do that, and at least one member of your homestead is acting extroverted, eventually word of mouth will provide you with plenty of customers, especially if your products are high quality.

The name game

When it comes to naming your homestead, you can get as complicated and esoteric as you want, or you can just name it after yourself or your family. It can be silly, it can be serious, it can be meaningful, or you might just like the way it sounds. Just make sure there's not another farm, homestead, or business engaged in a similar enterprise with the same name. If there's already a Jones Family Farm in your area, you're not going to make your homestead business stand out by calling it the Jones Family Homestead.

Social media and beyond

While the basics of marketing are the key, you will also want to consider putting together a website and establishing a social media presence. While a bit time-consuming and expensive, a website provides a way for your customers to find out more about you, as does your social media presence. In reality, your homesteading business is selling you and your unique approach to business. Let's be honest, your customers can buy a tomato anywhere, but they cannot support a unique and hardworking individual like yourself by shopping at a big box grocery store. So let your customers get to know you.

Your website can be as simple or complicated as you'd like—and can afford—but it should include some basic information:

Contact information—Your phone number, physical address, and email address. Your customers will want to call and email you about your operation and to find out what products you have available. While sometimes it can be time consuming, always take the time to be polite and open with your customers, whether they're visiting, calling, or otherwise reaching out to you.

Hours of operation—These aren't the hours you're working, but the hours you're willing to see members of the public. Remember, you want to see potential customers and make yourself available to them, but you don't want to do it to the detriment of the work on your homestead. Consider changing your hours of operation based on the season, making them shorter during the busy harvest and planting seasons and longer in the "off-season," when you might have more time to meet with customers.

Market availability—If you're going to be at a farmers' market, CSA drop-off point, or anywhere your products will be sold (restaurants, groceries stores, etc.) put it on your website. You'll want your customers to be able to find your products, wherever they may be. However, if you are not going to be consistently at any location, do not to list it. If you are trying out a market and only expect to be there once or twice during a season, don't list it as a regular spot on your website. Customers who go to a market or location and don't find you will be frustrated and eventually stop coming to see you, no matter where you are.

Product availability—You don't have to list every tomato variety or vegetable cultivar on your website (even though it's a good idea to do so), but you do want to let your customers know, at least in broad strokes, what they can expect from you. If you sell mixed vegetables, eggs, honey, and beeswax candles, your customers should be able to find out on your website. Update this section immediately if there's ever any additions or changes to the list.

Pictures of you and your operation—Again, homesteading businesses thrive on personal connections with their customers. Your customers want to know who you are and more about what you do. Images of both things help your customers feel like they know you.

Growing methods—If you're an organic or conventional farmer, tell your customers. And don't sugarcoat it. List any certifications your operation may have as well as information about how you produce your products and what's important to you. Your customers want to know how your products are grown and produced. Then they can make informed decisions about their purchases. Don't worry so much about being organically certified. Just explain to your customers, in as simple terms as possible, how you produce your products and why you use the methods you do.

For example, if you grow radishes, but you spray your radishes with a pesticide, explain on your website, that you have battled insect infestations on your radishes for years, and the only way you can get those vegetables to market is by using that particular pesticide. If your customers know the reasoning behind your decisions, they're more likely to understand and appreciate them. However, don't lie to your customers. If you grow food conventionally, don't tell them you use organic methods. Lying to your customers is unethical, and the damage it can cause when you're found out—and you will be found out—isn't worth it.

In addition to having a website, you will need to establish a social media presence. Facebook and Twitter are the bare minimums, but you should also be willing to think outside the box. Here's a brief list of some of the social media outlets you should know about and why.

Facebook—Facebook is intuitive, flexible, and ubiquitous. This social media behemoth boosts more participants than any other social media outlet, which means you can avoid it only to your detriment. The great thing about Facebook is its wide variety of options to engage your audience and customers. You can post photos, galleries, videos, notes, product lists, and more every day. The biggest tip regarding Facebook, or any social media site for that matter, is consistency and engagement. Post regularly and respond as quickly as possible to any requests for information, criticisms, or questions. Just remember, Facebook is a public forum, so act on your page as you would in any capacity in which you represent your business. Be courteous, polite, and helpful.

Twitter—Much more concise than Facebook, Twitter's 140-character limit is actually more of a strength than a weakness. While Facebook is powerful and all-encompassing, Twitter boasts nearly as many users, but with a quicker interface, both in terms of use and interaction. That means you can disseminate useful information to your followers faster. So, whenever you have a sale, a market cancellation, or any other information your customers need to know, Twitter is the go-to spot.

Pinterest—Pinterest is the scrapbook of social media. This service allows you to share and store articles and points of interest. This is a great tool for teaching your customers more about you and what's important to you. Favor biodynamic agriculture? Pin news articles about the movement to your boards. Want to discuss the difference between heirloom varieties of sweet potatoes? Pin a list article from another site. You can also pin blog posts and image galleries from your own website, allowing you a chance to brag about you, your products, and your homestead, further connecting you to your customers.

Etsy—More of a sales point than a social media site, Etsy still lets you connect to your customers, and provides you a chance to increase your overall sales and profit margins. Better for handicrafts and DIY projects than food products, Etsy is the craft fair of the Internet. You can use this site to sell your fiber arts (including raw fleece) and anything else you happen to produce. This site also serves as a showcase for your products. So, make sure to put your best foot forward by showcasing attractive images of your products for everyone to see.

Ravelry—One of my favorite social media sites, this quirky, esoteric corner of the Internet is best described as the Facebook for crocheters and knitters. Ravelry allows you a chance to connect with other fiber artists, who can offer you unique ideas, patterns, and marketing opportunities. While you can advertise your wares on Ravelry—any fleece or yarn you produce would probably be of value to others on the site—it's better to think of Ravelry as a brain trust in all aspects of the fiber arts. So you can reach out to other crafters and find out how they're selling their wares, as well as tips to keep your costs down.

Local avenues

Another often overlooked marketing opportunity occurs at the point of sale. Wherever you sell your goods and services, be it at the farmers' market, local restaurants, craft fairs, or elsewhere, you should take every opportunity to let your customers know exactly where their goods are coming from and how to get more of them. This can be done with signage, but look for other opportunities as well.

If you sell at restaurants, ask the owners if they can put information about your products on their menu (For example: *Bob's Burger's uses tomatoes produced locally by The Jones Family Farmstead*). If they can't (menus are often printed well before produce vendors are determined), ask if they can add your homestead business the next time menus are printed. In the meantime, ask them if you can put a flyer up in the window, on their door, or on their bulletin board.

Do the same at grocery stores by creating small signs that identify where the produce comes from. For handicrafts, make labels to let your customers know more about your operation. And, if you have a CSA, include a flyer in your CSA boxes each week. Sure, the CSA shareholder

Take every opportunity to let your customers know where their goods are coming from and how to get more.

already knows where they get their CSA box, but their friends and neighbors won't always know, so tucking a flyer or brochure in the box will give your shareholders a chance to let others know about it. So, if your shareholders are ever asked about where they get their delicious butter beans, they can just hand over the flyer. Whenever possible, make sure your signs, flyers, and other marketing materials reference your website. While you might not be able to spread the word about everything you do and sell in the limited space of a flyer, your web address will let people know where they can go to find out more about you and your products.

4 Section Four: Learning to Love the Fiber Arts!

CHAPTER 12: THE IMPORTANCE OF FIBER

As a step toward self-sufficiency

I love the fiber arts, and I'm convinced that true self-sufficiency can't be achieved without some sort of fiber component to a homestead.

I developed this attitude in a weird, roundabout way. About four years ago, my wife decided to take up crochet. She bought some hooks, yarn, and an instruction book, and she got cracking. Unfortunately, she didn't like crochet.

I, on the other hand, loved it.

I took to it like a fish to water and started crocheting everything I could, everywhere I could.

It's odd, but I discovered that people don't really expect a man to crochet.

Once, while visiting a relative in the hospital during surgery, my family and I were sitting in the waiting room, and I was crocheting a red scarf for my daughter. Hospital waiting rooms are perfect places to crochet; it gives you something to concentrate on to keep you from worrying.

My brother, who was sitting next to me staring at his phone, suddenly looked up after I'd gotten about a third of the scarf done.

Startled, he jumped up and walked off. When I asked him why, he expressed dismay at my newfound hobby.

"I don't think I want people to know you're my brother, sitting over there with your crochet!" he said.

I laughed at him. After that, however, I was visited by about four nurses, two doctors, and an orderly (all women, if that makes a difference). Each time I had the same conversation.

"I've never seen a man crochet," they'd exclaim.

"Yep," I'd respond.

"You're really good at it, too," they'd marvel.

"Thanks," I'd say. They'd walk off, shaking their heads.

Evidently, I was the talk of the hospital.

Since then, the fiber arts have become more popular than ever. Stitch 'n bitch groups have popped up all over the place, graffiti artists have started "yarn bombing," and some groups are even promoting the practice to help soldiers recover from PTSD.

Fiber arts are able to be a lot of things to a lot of people, but they have special significance for homesteaders. By practicing the fiber arts, you're taking yet another large step outside of consumerist culture. You're also producing extra value for your homestead, your family, and your pocketbook.

Fiber arts include:

- Knitting
- Sewing
- Spinning
- Crochet
- Needle Felting
- Needlepoint
- Macramé
- And more

Producing items from fiber will give you something to do on rainy days or during cold months, and they're great for bartering.

Essentially, by learning one or more of the fiber arts, you can monetize your homestead business more efficiently. You won't always be able to raise a crop or even gather eggs, but you'll always be able to make something beautiful.

Fiber arts are also a major step away from manufactured goods and toward self-sufficiency. Why buy a pair of mittens when you can make them? Why buy a dress when you can sew it? While fiber arts are deceptively time consuming, they can be inexpensive in terms of materials. For example, a skein of yarn can be purchased on sale for around four to six dollars, which you can then turn into a hat for your brother's Christmas present (yes, the very same brother who didn't want people to know that I crochet). While you might spend hours making it, those may be hours you'd normally spend watching television, which you can do at the same time . . . win! For those of you who are raising animals for their fiber (more on that in the next chapter), you may put in more time by making the yarn yourself, but it will cost you even less, and the yarn will be of higher quality.

As a commercial venture

In addition to saving you money and letting you step away from consumer culture, fiber arts are a great way to add value to your homestead. Anything you produce can be sold at craft fairs or on Etsy, and they can provide you with a lot of cash for your operation. Quilts can sell for hundreds, maybe even thousands, of dollars depending on the market and the designs. I know many needle felters who produce sculptures that are sold in museums all over the country.

The fiber arts are at their heart folk arts, which are becoming more popular with art critics and collectors all over the world. Selling your handicrafts at craft fairs, the farmers' market, and art galleries sets up an excellent tool for marketing. While selling your crafts at venues outside of the normal farmers' market, you will meet new individuals who are potential customers. A surprisingly large portion of your fiber arts customers will be the exact same people interested in your produce, honey, meat, eggs, etc. Remember, every interaction is a chance to expand your homesteading business, no matter how unlikely it may seem.

As an alternative to meat animals

Finally, my favorite thing about the fiber arts are the animals. I like animals. I don't necessarily like slaughtering and butchering animals (my wife is in charge of that task). But, as a pragmatist, I won't keep an animal that doesn't provide some sort of value to the homestead. So, I won't have a hutch filled with pet rabbits. Having too many animals as pets ensures you won't provide for them. There will also be a cost

The super soft fur of angora rabbits can be spun into yarn.

to feed and care for them, which means eventually you will not be able to afford to keep them.

So, if you are like me and dislike slaughtering animals—and you don't have a bloodthirsty spouse willing to do it for you—then you need another reason to keep them that will help offset the cost of caring for them.

Fiber is that reason.

For many meat animals, there is a breed that produces fiber—sheep, angora rabbits, alpacas, angora goats, cashmere goats—even the hair from Highland Cattle can be spun into yarn.

So, you can raise animals on your farm solely for their fiber production. And, if you put a little effort and planning into it, you can raise multipurpose animals. Think milking Highland cattle or angora goats.

Animals have a definite place on a self-sufficient farm; they're great at providing you manure to amend your soil, but if you don't have a way to feed them consistently and sustainably, then you won't be able to keep them long. Using animals for fiber allows you to avoid the icky aspects of ethical meat consumption, but allows you to still steward the animal in such a way as to serve your homestead.

CHAPTER 13: FIBER ANIMALS

Sheep, goats, rabbits, and alpacas!

Now that you're thoroughly convinced of the benefits of fiber animals, it's time to figure out how to get your full-blown fiber operation into production.

If you decide to raise fiber animals, you have four choices (notwithstanding the Highland cattle, which take a lot more space than most homesteaders have): sheep, angora and cashmere goats, angora rabbits, and alpacas.

Sheep, goats, and alpacas need a bit of room to move around, and you should have at least two of each. They are all social animals, so it's a bit cruel to raise just one. They get lonely if left on their own too long, which can lead to unhealthy behaviors that eventually cause physical problems.

The only exception I know to this rule, and I am more than a bit skeptical about it, is integrating an alpaca into a pre-existing herd as a guard animal. I think perhaps it might be better for the alpaca in question if she has a friend to help her guard other herds.

Rabbits also require friends, as they are also social animals, but they require much less space.

Also, another plug for rabbits: they produce my favorite poop. Rabbit poop is great for gardens and is considered "cool," which means it doesn't have to be composted before putting it on your garden, unlike "hot" manures.

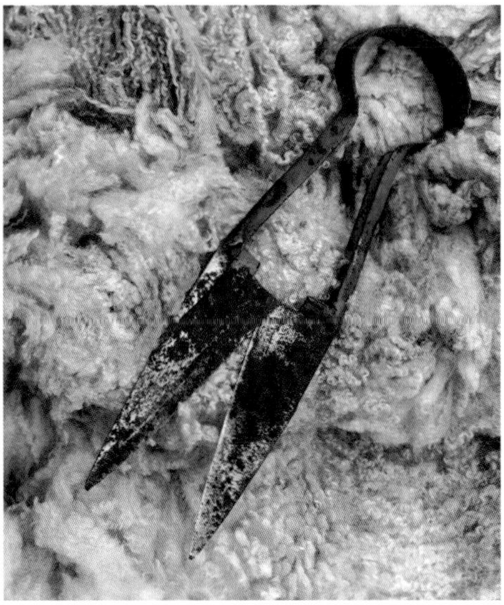

Shearing is an art form that takes much practice. If this is a step you'd rather skip, consider raising rabbits for fiber.

Wool gathering

Sheep, goats, and alpacas need to be sheared. While homesteaders can learn the art of shearing, it's something that takes practice to be able to do easily and efficiently, which benefits both you and the animal. Shearing can be done with machine shears or bladed shears. Machine shears are quicker, but bladed shears leave some wool on the animals. Using bladed shears might be a better option for homesteaders, as chances are you will be shearing fewer animals.

The bladed shears also leave more of the fleece on the animal, helping to protect it from the elements while the fleece regrows.

Highland cattle aren't sheared. Instead, the fleece is brushed off their coats.

Rabbits can be sheared, but the best way to get the fleece from them is with an air compressor or blow-dryer. The animals are placed on a tarp or sheet and blown. The loose wool flies off them and is collected on the sheet/tarp.

From fleece to yarn

Once you get the wool from your animals, you have to process it. The basic steps are cleaning, picking, dyeing, carding, and, finally, spinning. But you don't actually have to do any of these things. Many spinners and fiber artists actually enjoy the process of making yarn from start to finish. Additionally, it's also possible to sell your fleece to a larger operation. If you decide to do this, the buyer will tell you what steps she'd like you to take to before she gets the fleece.

If you decide you want to process your fleece, here are the basic steps you will need to take.

Turning the fiber from your animals into yarn requires several steps, but you may find that you enjoy it.

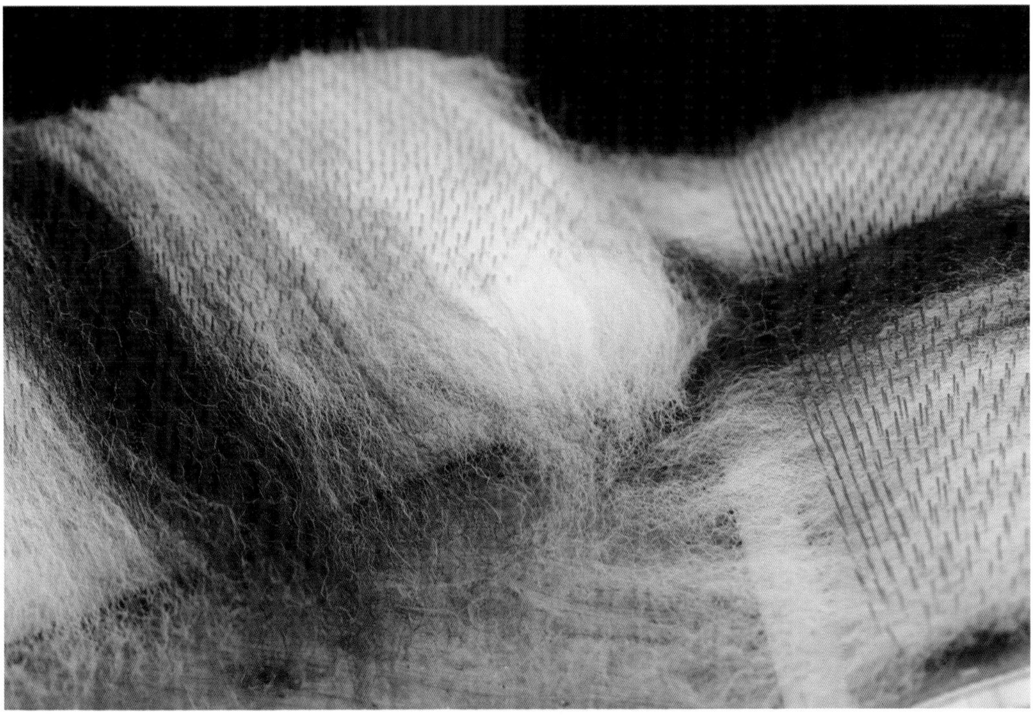

Once you've carded the fleece, you have batt and can start spinning yarn.

Cleaning—Animals are dirty. The fiber you get from those animals is also dirty. You will have to clean the fiber. This can be done by soaking it in a tub of hot water and detergent, or you can wash the fleece in mesh bags in the washing machine. Either way, the trick here is to remember not to agitate the fleece, it could lead to felting (felting occurs when the scales of the fiber lock together, creating … well, felt. It happens when wet fleece is pounded or agitated). If you use a washing machine, you'll let the drum fill with water, let it soak for about an hour, and then, drain, rinse, and spin it. However, make sure the material isn't agitated.

Picking—Even after cleaning your fleece, there will still be vegetation and little sticks in the fiber that you'll need to pick out. You can do this with a picker, which is a machine with what appears to be two beds of nails that work in tandem against each other. You can also just lay the fleece on a sheet and pick through it by hand.

Dyeing—I'm not going to discuss dyeing fleece at length here, as it's a complicated process that varies based on the type of dye used and the type of fleece used; for example, angora rabbit fleece tends to come out lighter after dyeing. Fiber can also be dyed after making yarn. I will, however, suggest a great resource to get started on dyeing wool. Chris McLaughlin, an angora rabbit keeper and master gardener based in California, wrote an excellent book, *A Garden to Dye For*. This book tells you how to grow plants that you can use to make your own dye. She also covers the basic steps of how to use those materials to dye your fleece.

Carding—Carding the fleece lines up the fibers so they're all set roughly parallel for spinning. Once this step is complete, you no longer have fleece, instead you have a batt. Which is pretty much the same as fleece, it just looks neater. Wood is put on hand cards, which have pins. The hand cards are dragged against each other in opposite directions. A drum carder can also be used. It works in exactly the same way, but uses two drums covered in metal pins spinning in opposite directions and at different speeds. Once you're done carding your fleece, you officially have batt and can, finally, start spinning yarn.

Spinning—Spinning can be done by hand (which I don't recommend), with a spinning wheel, or with a drop spindle. The process of spinning wool into yarn is mechanically simple. The point is to twist the fibers so they interlock in such a way as to form yarn. Honestly, I don't know much about using a spinning wheel, as I've never had enough wool to invest in one, but I can recommend a drop spindle. Drop spindles are simple devices with a storied tradition. The process can be learned by nearly anyone. And, since they're so simple, they can literally be made with an old compact disc and a wooden dowel. To learn more about drop spindles, check out *Respect the Spindle* by Abby Franquemont.

Homesteaders' favorite fiber animals

Rabbits

- Angora: English, French, German, Giant, Satin—The fiber from angora rabbits can be harvested with minimal equipment and produces an excellent and popular wool. The wool is sold by cottage industries as a fleece, spun into yarn by hobbyists for crochet and knitting, and any operation can be scaled up to produce a sizeable commercial venture.

Goats

- Angora—Angora goats originate from the Middle East and are known for their high-quality fiber production. Varieties have been bred all over the world with minor differences being produced due to the breeding practices of the various countries in which they are raised. The goats produce mohair and look like goats crossed with sheep due to their fluffy appearance.
- Australian cashmere—This goat is popular in Australia, where it is known for its high-quality midwinter coat. The breed has its origins in the 1800s, when a breeder encouraged the country to support a fleece industry. The goats were kept for a time until abandoned in the early 1900s (due to a gold rush), only to be rediscovered and revived as a breed in the 1970s.

Sheep

- Booroola Merino—Originating in Australia, this breed is known to be small. The size, however, is offset by the breed's docile nature and fine wool. This size also makes it suitable for small farm operations. Although it originates in Australia, the breed has spread to New Zealand and the United States.
- Delaine Merino—One of the oldest sheep breeds in the world, it is prized for its long wool and high production rate. Folds in the breed's skin tend to produce more surface area for wool to grow on. They are hardy and bred to thrive in the Southwest region of the United States.

For those who want to raise animals for their fleece, alpacas are another option.

- Cotswold—Cotswold sheep are an English breed, originating in the Cotswold Hills area of Gloucester. They are dual purpose, raised for both meat and wool production. The breed is considered well suited to shepherds, as they are known for their calm demeanor. Another old English breed, there are some sources that say they have been in the region since Caesar's visit to the United Kingdom. They have been renown through the centuries for their fine wool, while more recent breeding has led to their dual use as a meat animal.
- English Leicester—Another English breed, this sheep is known for its excellent long wool. While the breed is now rare in the United Kingdom, it is one of the few truly purebred lines of sheep. It is known for its wool production, and recent breeding innovations have led to the breed's use as a meat animal.

CHAPTER 14: CROCHET, KNITTING, MACRAMÉ, SEWING, AND MORE

Fiber arts basics

I love crochet. It's my go-to fiber art. And, as previously mentioned, I've "suffered" from the good-natured teasing from my brothers and sisters about my passion. But it's led to some really great stuff.

Last Christmas, my sister bought me a series of books about nearly every fiber art under the sun. Originally published in the 1970s, the books provide patterns and methods for just about every means of putting string together in order to make stuff. I believe she intended for the books to be a joke gift, but, honestly, they are now among the prize tomes in my little library.

Since you now have yarn that you made from your very own drop spindle—or yarn that you bought, no judgment—it's time to actually make something with it.

Assuming you haven't chosen your favorite fiber art, here's a brief description of several of the fiber arts. You can pick one that sounds fun, or you can pick several, it's up to you.

Crochet—As previously established, crochet is the best of all fiber arts. Originating in Europe in the 1600s, crochet was first used as a method to craft lace using tiny hooks. It wasn't until the 1970s, when crafters started

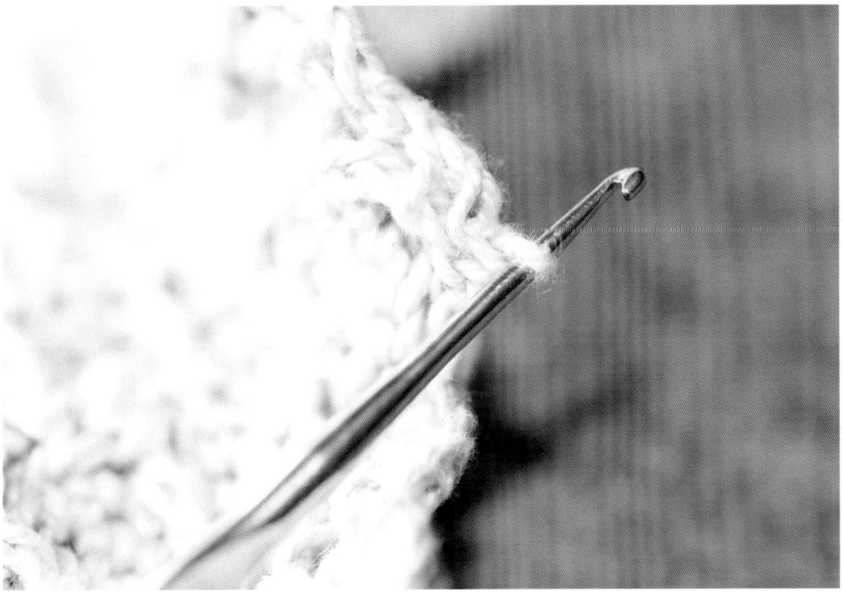

using larger hooks to create fabric with the same methods, that crochet became the art form it is today. In recent years, crochet has become even more popular due to the Japanese derived art of amigurumi, wherein small, sometimes intricate, and decidedly cute stuffed animals are created by crocheting and knitting. They're usually considered more rugged when knitted, as knitting tends to produce tighter stitches. With the proper methods, however, there's no reason crochet can't be used to create the same items and cloth as knitting.

Knitting—Knitting is the most popular of the fiber arts. Just about any garment can be knitted. A pair of cotton socks found in Egypt from the first millennium AD are one of the earliest known examples of knitting. Knitting is most commonly seen being done with a pair of smooth, relatively long needles made of wood, metal, or plastic. However, it can also be done with circular needles, which consist of two shorter tapered needles that are connected by a flexible cable, or a set of double-point needles. Early knitting needles may have been made from bone, ivory, or even tortoise shell.

Sewing—The act of joining cloth together with thread, sewing is versatile and handy. Although primarily done for making clothes, a dedicated seamstress can make just about anything: clothes, blankets, aprons, even tents. While sewing is a part of quilting, quilting is usually considered a completely separate art.

Quilting—The creation of quilts and other large items using patchwork designs. When making quilts, batting is sandwiched between two layers to provide

insulation. The art form is varied and expressive. Either by hand or machine, a quilter can create folk-inspired designs, often with a rich and storied history among regions and families, as well as more modern designs that may resemble mosaic or modern art. In fact, quilts are often displayed at galleries and museums for their artistic merit and historical significance.

Needle felting—Needle felting is the use of a barbed needle to felt fleece roving to create three-dimensional shapes. Sculptures, wall hangings, dryer balls, and even pillows can be made from needle felting. By using different colored wool in the process, startlingly complex designs and colors can be produced.

Macramé—Invented by prehistoric man as a method of making cloth without weaving, wall paintings from ancient Egypt show people wearing garments made with macramé. Usually consisting of a few knots repeated over and over again, macramé has the advantage of being wildly versatile. Necklaces, hammocks, curtains, nets, bags, and more can all be made with macramé methods.

Outlets for your goods

Once you decide on a fiber art and start creating, you'll need to do something with the goods you produce. While many of the items you create will be for your personal use or for your friends and family, you cannot overlook the commercial potential for these items.

With that in mind, once you get a large enough stock of items—a couple dozen pieces, at least—you should look for venues to sell your work.

Craft fairs are an obvious outlet, and some farmers' markets allow for the sale of crafts. If you're already selling vegetables, eggs, meat, or other goods at a market, you might want to consider expanding your offerings there.

You can also look into boutiques. If your work is good enough—and I promise, with enough practice and passion, it will be—specialty boutiques will be happy to have it. As with any form of sales, make sure your customers know about you and your work. Fiber arts, unlike other homestead offerings, can be commissioned, and commissions often cost a premium.

Don't forget the previously mentioned museums and galleries. There's no reason not to consider these works as a fully artistic expression of the maker. Treated as such, a macramé wall hanging, quilt, or crocheted amigurumi figure would fit right in at an art show, and may fetch a higher price than usual.

Bearing all those options in mind, don't forget the ubiquitous Etsy.

While Etsy provides a market for your work, it also gives you a chance to expand your reach for customers, as well as a method to connect with like-minded individuals.

Remember, however, that Etsy is just one of your market options. Other online marketplaces include Handmade Artists' Shop, AFTCRA, and Handmade at Amazon. Ask others which marketplaces they have found success with. You can also check online for reviews of the various sites.

Pricing your goods

When pricing your goods, make sure you charge enough to cover the cost of material, with at least a little for labor. Make sure, however, you at least check the local markets for similar products and price yours accordingly.

The longer you practice, the better your work will become, and the faster you can produce goods. After a while, you will have the knowledge and ability to establish a price based on what the market will bear.

5 | Section Five: Home Production

CHAPTER 15: CANNING

Lessons learned

The first successful harvest I had—and I use successful semi-ironically, as I spent more money than I made on the harvest—was from an excellent variety of hybrid squash I planted called Green Griller.

Green Griller is a dark green squash developed specifically to grill or roast. The plant produces round, slightly oblong fruit that's slightly bigger than a large man's fist.

It was delicious.

I planted nearly three hundred row feet. This variety produces a crop thirty-eight days after germination, which is really fast for summer squash.

So, I sold as much of the squash as I could, but one day I did the math and discovered I had grown more than six hundred pounds of squash.

I only sold half of it. I was drowning in squash. Absolutely drowning.

This is a common problem with cucurbit growers (squash is part of the cucurbit family). In fact, there's even a National Sneak Some Zucchini onto Your Neighbor's Porch Day (August 8).

Unfortunately, I didn't know about this problem among gourd growers.

So, I gave away as much as possible. I found needy families and all sorts of people. I made zucchini bread and fried and baked and ate squash like a crazy person.

At the end of it all, I had nearly two hundred pounds of squash on my kitchen counter and kitchen table; there were even tubs of it on my office floor.

At the time, I did not have a chest freezer. I did not have a pressure canner or a water bath canner.

If you are going to grow your own food, you better have a way to deal with it come harvest time, or you will throw a lot of it away.

After that, my wife and I took classes on canning, fermenting, and drying so we would never be in that situation again.

A bit of canning history

First, a little bit of history. Food preservation has always been sought by home cooks, small farmers, and homesteaders. Drying, fermenting, and root cellars are methods that have been attempted over the years. In fact, some believe pie was invented as a food preservation method, as people would cook stews and meat inside thick crusts to store them during the winter. Most of these methods worked, at least to some extent.

Some of these processes were time consuming and didn't produce large enough amounts of food to feed, say, an army.

So, during the Napoleonic Wars in Europe, the French government offered a prize to anyone who could derive a method of preserving large amounts of food for an extended time frame to feed the French Army, which was moving across Europe at the time.

Nicolas Appert, a brewer, noticed that food cooked in jars didn't spoil if the seals held. He developed a method of putting food inside modified wine jars, sealing them with corks, and boiling the jars. Appert didn't know it, but the process he stumbled upon was later codified and became pasteurization. Because the boiling killed harmful pathogens in the stew and the lack of oxygen prevented new ones from growing, the food stayed disease-free. He was awarded a prize of ₣12,000 (about $240,000 today) for his invention.

After the war, his method spread all over Europe. In England, because of glass shortages, tin cans were used and soldered closed with lead.

Eventually, the technique made its way to America, where glass jars had tin lids sealed with wax. The method wasn't reliable until John Landis Mason developed the mason jar, which featured a screw-on ring and a rubber-sealed top, in the late 1850s. The technology made canning a much more reliable activity and much safer as a result.

Mason's jar, tragically enough, never made him rich. Manufacturers like Ball and Kerr started manufacturing the jars after Mason's patent expired in 1879. Mason died in poverty in 1902.

Canning became a popular American pastime through World War I and II, as Americans and other countries faced food shortages. Home gardens, called Victory Gardens during World War II, became ways to supplement food for all Americans, and canning was an excellent way of preserving those harvests. During World War II, 41 percent of the food consumed by Americans was produced in those gardens.

Before World War II, as well as and during and after, canning was taught to American households by the USDA's Cooperative Extension office (established in 1914).

Through the years, Cooperative Extension developed the safety standards and safe food methods still taught in canning classes at extension offices all across the country.

Inspect your jars

Inspect your canning jars before using them. Since canning jars are glass, repeated use and damage during shipping can lead to fractures, cracks, and chipped lips. Not only will this cause improper sealing, but these damages can also lead to broken jars during the canning process, which can be dangerous. When a jar fails, which is almost always due to damage in the glass, it simply pops. (Note: Jars may explode, but it's incredibly rare.) Additionally, the mason jar canning system is designed in such a way that the lid will usually fail, releasing pressure, before the glass will fail.

Canning safety

Despite many first-time canner's fears, safe canning isn't all that hard. In fact, you can keep your family safe from food-borne illness most of the time by using the smell test: if an open can of food smells, or looks, funny, don't eat it. (Note: Botulism, which is the most dangerous food-borne illness that can crop up in canned foods, is tasteless, so make sure you follow the rest of the safety rules).

The best way to prevent food-borne illness during canning is to follow a few simple rules, the biggest one being wash everything. Wash your hands, wash the utensils you'll be using, wash the jars, and wash your hands. (I know I said it twice; it's that important.)

Here are the basic safety rules you should be following:

- Wash your hands (really, it's crucial).
- Wash your utensils, jars, equipment, etc., anything your food might remotely touch, wash it.
- Don't water bath can (more on that) any foods that are considered low acid. Low acid foods and all meats, including fish, should be canned in a pressure canner.
- Can your food as quickly after harvesting as possible.
- Do not reuse lids. Canning jar lids are single-use only. Lids that have been used before may have cracks in the seals or be rusted through, letting oxygen, bacteria, and mold spores inside the can.
- Do your research. Different foods require different cooking temperatures

to kill bacteria. Get the USDA canning guide to learn more about the specific temperatures and pressures needed, available free as a pdf, at: http://nchfp.uga.edu/publications/publications_usda.html.

> **Check the recipes and specs first!**
>
> Different foods have different cooking and canning times, no matter which canning method is used. These cooking and canning times have been researched by the USDA for optimum freshness and safety. So take the time to look it up before canning. It won't take long, and it can definitely save you trouble—those times are designed to increase the odds of sealed jars!

High Acid Foods	**Low Acid Foods**
Citrus fruits	Green beans
Apples	Corn
Pears	Carrots
Tomatoes	Onions
Blueberries	Peas
Cherries	Asparagus
Strawberries	Squash
Peaches	
Pickles	

Water bath canning

Water bath canning is the simplest, and probably cheapest, form of canning as far as investment in equipment goes. A water bath canner is essentially a stockpot with a lid and an inside rack that holds the jars. New water bath canners are not expensive, but I've bought about three of them for a few dollars each at yard sales and thrift stores. Most of the thrift store finds don't have the inside rack, but you can make do with spare canning jar rings lining the bottom.

A water bath canner is good for processing high acid food—pressure canners can be used, but they take a bit longer.

First, clean all the equipment and jars. Use a clean jar lifter after you clean your jars to handle them to prevent contamination from your hands (even though they should be clean, right). Next, pack the jars with food. While you can raw pack your food (pouring boiling canning solution over the raw food packed in the jars), I recommend that you hot pack your

> **A note on green beans**
>
> Many people, myself included, have memories of their grandparents canning green beans with a water bath canner. Green beans are low acid foods and should be canned with a pressure canner. Yes, my sainted Granny canned green beans in a water bath canner, and I've eaten countless jars of green beans from her stores. I didn't get sick because I got lucky, she got lucky, and all the other people who did it before us got lucky. Some of them did not, however, and improper canning has led to sickness. Instead of risking it, just follow the safety instructions.

Canning jar rack hack

To make a "rack" for your water bath canner using canning rings, just place clean (sterilizing isn't required) canning rings, with the small side of the ring up, on the bottom of your canner. Cover the base of the pan with as many rings as you can fit, packing them in as tightly as you can. If you still have some looseness in your rings, don't worry; the jars will sit on top of the rings. The whole point is to prevent the jars from touching the bottom of the pan, and the rings do a fine job of this.

food. Hot packing involves cooking your food for several minutes in the canning solution before putting it in the jars. This reduces air in the food, and therefore in your jar, which helps prevent bacteria growth and food shrinkage.

After packing your jars, you'll use a special tool, called a bubble remover, to remove air bubbles by putting it all the way into the jar and then shaking the jar. Once the bubbles have been removed, put a lid on the jar. (Magnetic lid holders are available and prevent you from touching the lids with your hands.)

Screw a jar ring onto the jar tightly to hold the lid down while the canning seal is set.

Finally, using a jar lifter, place the jars into the canner, and get the water boiling. As noted previously, different foods need different boiling times, so check a reference book to ensure you don't overcook or undercook it.

Once the jars are done, pull them out of the canner with the jar lifter and put them on the counter to cool.

As the jars cool, you might hear a ping or pop as the lids seal. It's a lot of fun!

After cooling, check the jars to make sure the seal is set. Push on the tops of the jars and check for give. If the lid moves up and down, the seal did not set properly.

If it isn't set, you can eat the food within a few days (store it in the fridge) or reprocess the contents entirely. Be sure to throw away the faulty lid.

Once your jars are sealed, you can remove the rings and use them for other jars. The lids will remain on the jars without the ring.

Pressure canning

Pressure canning is a bit more complicated than water bath canning. Pressure canners have devices, either counterweights or gauges, that tell you how much pressure is inside the canner. Counterweights prevent steam from leaving the canner until a certain pressure is achieved. Gauges show you the internal pressure of the canner on the outside. Some gauges offer users the ability to set a pressure. Once that pressure is achieved, the canner emits steam. Pressure canners must be sealed when they are in use, and they must be vented of excess air before using.

Essentially, when you put your jars in a pressure canner along with a predetermined amount of water and seal it, there is still some air inside the canner. The air expands at a different rate than the steam from the water, giving a false pressure reading, which can lead to faulty seals on your jars. Once you put heat to your pressure canner, the water inside will boil until it becomes steam. This steam will push air out of the canner through the vent port (the part of the canner lid where the pressure indicator sits).

After ten minutes of steaming, the air is completely out of the canner. Then the pressure indicator is put back on the vent port. When the pressure is high enough, remove the canner from the heat.

Once it cools, open the vent port and release the pressure. Once the pressure is *completely* relieved, remove the lid, take the jars out with a jar lifter, then let them cool and seal.

Do not force-cool your pressure canner; some home canners will run cool water over a hot pressure canner. Not only can this cause broken and damaged seals, it can also cause cracks in the canner, which can lead to a dangerous release of steam. Check the owner's manual of your pressure canner for more safety information.

Again, test the seals with the method listed for water bath canners. Remove the rings and store the jars.

Pressure canner replacement parts

Cooperative extension offices sell seals and other parts for pressure canners. If your local extension office does not have replacement parts, they can direct you to stores that sell replacement parts. Designated cooperative extension agents can also inspect your pressure canner for cracks, damaged seals, and safety. Most cooperative extension offices provide this service for free or for a small fee. For your safety and for better results when canning, it's a good idea to have your pressure canner inspected at least once a year.

Storing canned goods

Jars should be stored in a dark place that will not reach temperatures higher than ninety-five degrees Fahrenheit.

Jars kept in this manner can be kept for at least a year without impacting food quality or nutrition. Lids should be marked with the canning date (stickers, sharpies, any method).

It's also a good idea to label the contents of jars, especially those containing meats, stews, soups, jellies, jams, and preserves, any food that cannot be identified at a glance.

Jars can be stacked, but it does increase the risk of damage to the lids. Any jars with damaged lids, broken seals, or cracked glass should be emptied and discarded. Jars with mold or visible growth should be discarded. Jars with rusted lids should be carefully inspected for leaks or broken seals before consuming.

If you ever have any doubt about the safety of your food, discard it.

Top five canning mistakes

Have you ever gotten your hand on a perfect crop of green beans, decided to can them, then spent hours of work dealing with jars and lids only to have it turn into a big soggy, cloudy mess?

So have we. We asked Robin Seitz, a canning expert and former extension agent with the North Carolina State University

Be sure to label and date the contents of your jars, then store them where they will not be exposed to light.

Onslow County Center, about the most common mistakes she's seen in home canning.

As the go-to guru for canning in Onslow County, she fielded calls all season long and helped people work through common issues. Here are her top five canning mistakes from her years of experience.

1. Processing low acid foods in a boiling water bath

This is the most common error Robin sees. Home canners will use a water bath canner and try to can low acid foods. This is a problem. Not only can you wind up with a soggy, unappetizing mess instead of pristine jars of delicious food, you can actually cause undue harm.

"The temperature [in a water bath] doesn't get high enough to kill the bacteria you're trying to kill," Robin explained.

Sure, you may have done it. Sure, it may not have killed you, but that just means you got lucky. Try not to tempt fate again.

2. Being in a hurry

If you've spent hours—sometimes days or even weeks—canning the latest crop of potatoes from your garden (or your local farmers' market), then you're probably guilty of this one.

You look at all the time you have left for that pressure canner to cool, and then decide to speed up the process since you've got things to do. Maybe you have to pick up the kids from school. Maybe you have to sew a Halloween costume. Maybe you volunteered to help at your church's bake sale.

So you take the canner and drop it into an ice bath or put it out on the porch so the wind will cool it down a few minutes faster. What's the harm?

Plenty. Cooling your product down too fast results in a rapid pressure drop, which messes with the quality of your jars of potatoes (or carrots, or whatever).

Robin said, "If you don't let the pressure drop naturally, you'll get evaporation."

That means you'll lose liquid in the jar, and instead of full jars of delicious potatoes, you'll have some jars (and sometimes a lot of them) that are suddenly half full of product, exposing your veggies to unwanted air.

While it isn't a safety issue—pressure canners do a great job of killing bacteria—it means the results will be less than appetizing.

So take your time and work on the sewing projects in between batches. It's worth the effort.

3. Not taking out the air bubbles

There's a little device that comes with canning tool sets. It's usually just a simple plastic stick, but it can make all the difference. When dealing with chunky vegetables, large air pockets can get stuck at the bottom of the jars. You think "No harm done, that'll come out in the process."

And it does. But those air pockets can be quite large, which means when you pull those jars out of the pressure canner, those bubbles have risen to the top, resulting in a big drop in liquid levels.

Robin said, "You think, 'My liquid cooked away,' [but] it didn't."

To deal with this issue, take that little plastic stick (or a clean spoon, a butter knife, or a popsicle stick) and pop those suckers. It's kind of satisfying.

4. Packing the jars too tightly

At the end of every canning process, you'll have a handful of veggies that just won't quite fill up a jar. So you think, well, I'll just cram them in the top of another jar.

That's problematic.

The extra product takes up space. And while that half inch or so of empty space between the contents of a jar and a lid may not seem like a big deal, that space is crucial. It prevents moisture from degrading the seal of a jar. Pack in too many veggies and that seal can become compromised. That means in the wintertime, when you're looking forward to that nice jar of green beans in the cupboard, you get a mushy mess instead.

So take those extra green beans (or whatever) and throw them in a cast iron skillet with some olive oil and minced garlic (that you grew yourself) and have a little snack instead. You deserve it.

5. Reusing lids

With a few exceptions, canning lids are one and done. That means no matter how good a lid looks after using it, don't use it again.

"A lot of people don't realize not to do that," Robin noted.

The rubber seals on canning lids develop tears and fissures during the first use. Sometime the tears are microscopic, meaning the seal will look pristine to the naked eye, but still be unable to hold a seal. And sometimes the heat from the canning process can close the tears up, which means the jars will seal at first, but as soon as the jar cools enough it will open up, leading to air and bacteria exposure. So don't reuse your lids!

The most important tip from Robin is the simplest: "If it doesn't look right, if it doesn't smell right, don't eat it."

Happy Canning!

Be sure to avoid the top five canning mistakes when preserving the fruits of your harvest.

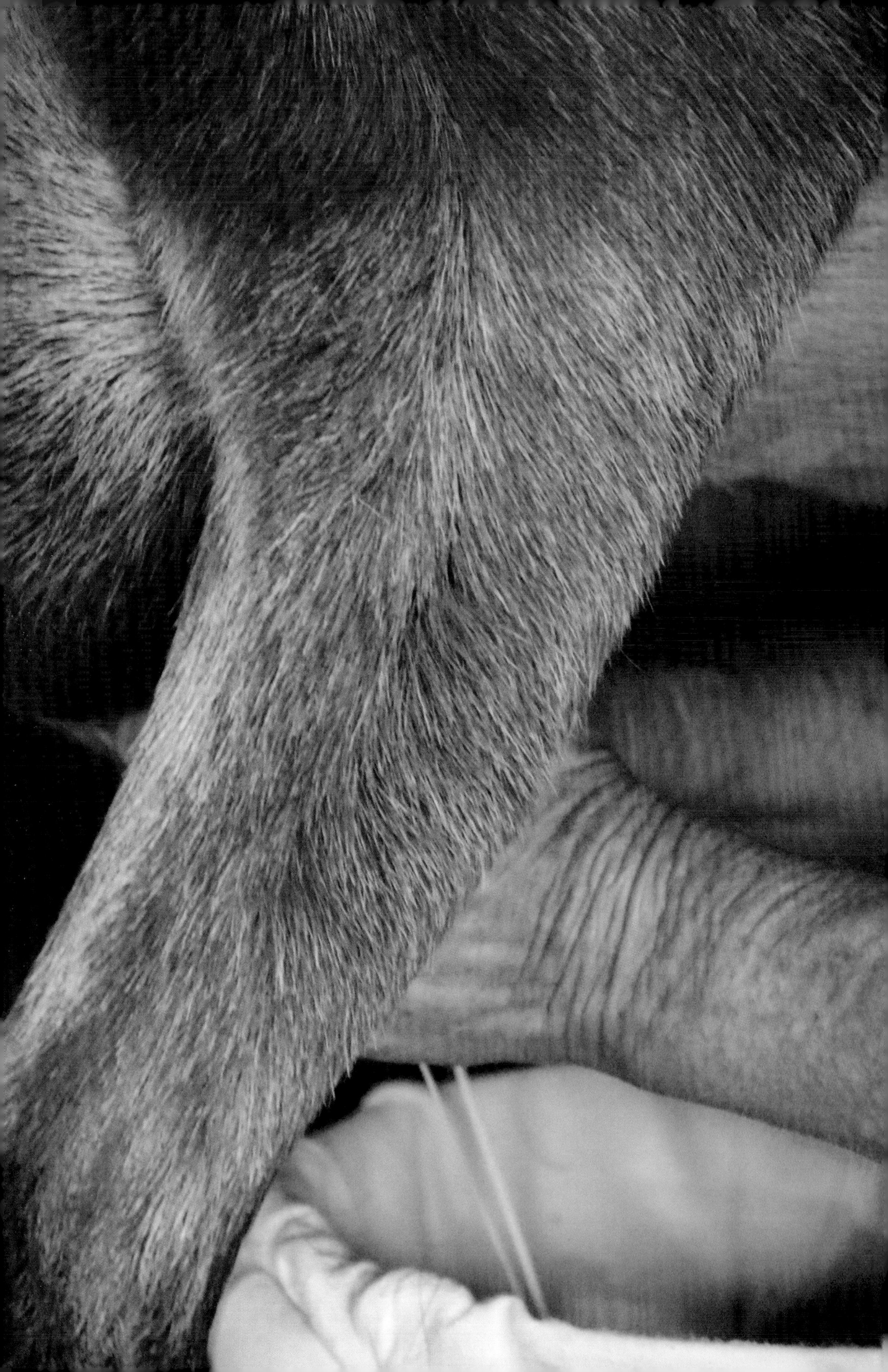

CHAPTER 16: DAIRY

Milk and more

When we first acquired some milk goats, we were under the impression that the one goat who was lactating at the time would barely provide enough milk for us to use.

We were very, very wrong.

That goat produced half a gallon of milk a day, sometimes more. She was a good producer of milk, but not at all unusual.

We drank as much milk as we could, but, honestly, our family is not that into drinking milk.

We cooked with it, added it to coffee, and eventually started freezing it (milk can also be canned).

After a time, we finally decided to start making cheese, butter, and yogurt. All three are more nutritious than milk alone and can be kept for longer periods of time without spoiling.

While we went in unprepared, dairy turned out to be a big piece of our little homesteading puzzle.

Dairy standards

Raw milk sales are heavily regulated in the United States. Raw milk refers to unpasteurized milk. If you decide you do want to start a dairy, even a small homestead dairy, you will have to purchase a pasteurizer and submit to an inspection. The regulations for a cheese-making dairy, a Grade B dairy, however, are much more obtainable than that of a Grade A dairy, which is required for actual milk sales. For more information on dairy standards, contact the USDA or your local Cooperative Extension office.

At the time, most of the protein we produced came from eggs and beans. (We were not culling chickens for meat, that came later.)

The caloric content of nearly everything we produced was relatively low. Most vegetables, for better or worse, do not have that many calories. Exceptions exist—potatoes, sweet potatoes, etc. —of course, but if your goal is to produce calories on your property that can significantly sustain life, then you need

The history of cheese

No one is entirely sure how or when humans started making cheese, but we do know that the production of cheese goes back more than seven thousand years. Cheese making equipment dating back to that time was discovered in Poland. A theory exists that the process was discovered when early dairy farmers and herders used cow stomachs to store and transport milk—a common practice at the time. The bacteria that naturally occurs in the stomachs of cows is perfect for cheese making, which led to the milk forming curds. Since cheese can be defined as curds plus salt, it's not that big of a leap to see how those early farmers started us on the path to cheesy goodness today.

something like dairy to provide those big calorie boosts.

For us, that was goat's milk. A gallon of goat's milk contains more than 2,500 calories.

The process of making cheese, yogurt, and butter concentrates the proteins, fats, and carbohydrates in milk, essentially by removing the water and using microbial and enzyme activity to make those calories more available.

Also, as mentioned, cheese, butter, and yogurt store a lot longer than raw milk. And, if you're able to sell your dairy products, these items fetch a premium, as small-batch, artisanal cheese, butter, and yogurt are still hard to get, even in this modern DIY culture.

If you don't have goats (or sheep or cows or even buffalo), you can still grab some milk and a few supplies and start learning about how your food is made while picking up valuable skills in the process.

Making cheese

The basics of cheese making are similar across the board. You take milk, add a catalyst to separate the milk, and then use live cultures to turn the curds into cheese.

Technically, there are a lot more steps, but the broad strokes are there. I know when I first started making cheese, I was intimidated, the steps seemed long and overly complicated, and I was constantly terrified of messing up somehow. Really and truly, cheese is a bit hard to mess up. If you have the proper ingredients, you will wind up with cheese.

To start, here are some basic terms to keep in mind.

Curds and whey—When milk is separated into its basic forms, you get curds and whey. Curds consist of the fats, proteins, and many of the sugars that make up milk. Whey contains a few odd nutrients, but it is mostly water. The process is essentially what happens when milk curdles in the refrigerator, but when making cheese, you control how it happens. The curds are used to make the actual cheese. The whey can be used as a nutritional supplement for your family or it can be fed to your animals. Chickens love it!

Rennet—Rennet is a group of enzymes produced naturally in the stomachs of ruminant animals: cows, sheep, goats, etc. These enzymes are crucial in the production of cheese, as they allow the milk to curdle and promote the growth of the flavor-producing bacteria in your cheeses.

Fresh and aged cheese—Fresh cheeses, such as mozzarella, cream cheese, and cottage cheese, aren't aged like cheddar or gouda. Fresh cheeses are usually easier to make, but differ greatly in texture and taste.

> **Good news for the lactose intolerant**
>
> Properly aged cheese does not contain lactose, which means it can be consumed by people who suffer from lactose intolerance. The aging process consumes all the lactose in the cheese, rendering it safe for sensitive stomachs.

Making yogurt

Yogurt as a milk product is tied directly into human civilization. Believed to have been produced in ancient Mesopotamia, yogurt has been a staple in diets all over the world ever since.

Oddly enough, yogurt wasn't popularized in the United States until the early 1900s, when Dr. John Harvey Kellogg popularized the product as a treatment for nearly every disorder known at the time. He was particularly fond of prescribing yogurt enemas.

While Kellogg was a bit of a kook by all accounts, it appears he may have been on to something. Modern research shows that the live bacteria in yogurt is beneficial to producing healthy fauna in the human digestive system.

Yogurt is really easy to make. You can purchase yogurt cultures or just buy the yogurt you'd like to make and use it as a starter. However, make sure the yogurt you use as a starter is marked as containing "live bacterial cultures."

Making butter

Butter is one of my favorite things to make, and you'll find out why shortly.

Butter is the concentrated fat of milk. While butter was consumed all over the Old Word, it was only really popular as a food staple in the northern climates, as butter quickly goes rancid in warmer weather.

Butter is made from the cream of milk, which is one of the reasons why cow's milk is so popular. Goat's milk is naturally homogenized, which means the butterfat must be separated before you can use it to make butter. Depending on your level of commitment to butter making, this may necessitate purchasing an expensive cream separator. You can however, separate the cream from goat's milk by putting it in a shallow pan in the refrigerator and letting the cream rise to the top over a few days. As it does so, you can skim the cream from the top of the pan with a large spoon and put it in a separate container. Once you get one or two cups of cream you can make butter.

The process of making butter from cream is called churning. The churning process agitates the milk. The membranes of the globules of butterfat in the milk break during the agitation, allowing the particles to join with other particles of butterfat. As the clumps of butterfat get bigger and bigger, they form butter.

After churning long enough, the butter is removed from the milk. The leftover milk, called buttermilk, can be consumed or used for cooking. The butter is then rinsed, packed into a butter mold, and chilled until stiff.

Farmhouse Cheddar

Here's a quick recipe for cheddar cheese that you can make at home.

Equipment

- cooking thermometer
- 16-quart stockpot with a double boiler long knife (any kind will do as long as it reaches the bottom of the pot)
- whisk
- cheesecloth
- cheese press (optional, but a really good idea)
- strainer

Ingredients

- 3 gallons of whole milk
- 2 teaspoons calcium chloride (only for pasteurized milk)
- mesophilic culture (¼ teaspoon ABIASA, ⅛ teaspoon Danisco, or 1/16 teaspoon Sacco)
- 1½ tablets rennet or ¾ teaspoon liquid rennet
- ¼ cup unchlorinated water
- 1 tablespoon salt

Directions

1. Put the milk (with the calcium chloride, if necessary) into the stockpot with a double boiler.
2. Slowly heat the milk to 86 degrees Fahrenheit. Turn off the heat, then stir in the culture. (Different types of culture create different flavors of cheese.) Stir constantly but gently until the culture is incorporated. Cover the mixture and allow it to rest, keeping the temperature at 86 degrees Fahrenheit for 45 minutes.
3. Dissolve your rennet in the water.
4. Keep the milk at 86 degrees Fahrenheit. Stir the rennet into the milk, then let the milk sit for about 30 minutes, until the curd forms completely.
5. Cut the curds into ½-inch-wide strips, then stir the curds to break them into chunks. Let it sit at 86 degrees Fahrenheit for another five minutes.
6. Slowly heat the mixture to 102 degrees Fahrenheit, about 8 minutes (ensure that it does not scorch). Stir the curds gently while heating. The curds will break up as you stir them, until they're about the size of shelled peanuts or even slightly smaller. Keep the curds at 102 degrees Fahrenheit for 30 minutes, then remove them from the heat.
7. Rinse the strainer in hot water to preheat it, then pour the curds and whey through the strainer into a pot or bowl.

Got goat milk?

Homogenized milk usually refers to milk that contains the unseparated butterfat. In the case of cow's milk, which requires homogenization, this means a portion of the separated butterfat (aka cream) must be stirred back into the milk before it can be called whole milk. Milk with 2 percent of the butterfat stirred back into it is 2 percent milk, whereas milk with all of the butterfat skimmed off the top is called skim milk.

Goat's milk contains less butterfat than cow's milk, but has the advantage that it doesn't have to be homogenized, as it's already "mixed" together when it leaves the goat's udder.

8. Put one-third of the whey back into the main pot along with the curds. Put a lid on the pot, then cover it with cheesecloth or a dish towel to keep it warm. While in the pot, your curds will lose even more whey. This is the cheddaring process. Let your curds cheddar for about 1 hour.
9. After cheddaring, you'll have a slab of curds. Cut them into small pieces with a knife.
10. Add the salt, then knead it into the curds.
11. Wrap the curds with cheesecloth and press in the cheese press until you get a flat disc. If you don't have a press, you can pack the curds in a piece of cheesecloth and squeeze the moisture out by pressing the mass of curds a drying rack. You want to remove moisture from the curds while at the same time pushing them together into a single mass.
12. Remove the disc from the press. Remove the cloth. Let the cheese sit on a wire rack for about 12 hours.
13. Wash the cheese with a mild saltwater brine (1 tablespoon salt to ½ cup water). Pat it dry.
14. Store the finished cheese away from direct sunlight and temperature extremes. Wrapped in plastic wrap, it should last for several months. For longer storage times, you can cover the cheese in wax and age it. Waxing and aging cheese is a complete process on its own. If you're interested in continuing to make cheese, I suggest you grab a copy of *Artisan Cheese Making at Home* by Mary Karlin.

Pasteurized or unpasteurized

Whether you use pasteurized or unpasteurized milk for cheese is up to you. Unpasteurized milk separates better, but it isn't allowed to be used in cheese that's sold to the public. Ultra-pasteurized milk is completely unsuitable for cheese making, as no amount of sodium chloride is going to make that material break into curds and whey.

Quick and Easy Yogurt

Yogurt is simple to make, amazingly good for you, and super tasty.

Equipment

- 2 clean quart jars with lids
- double boiler
- candy thermometer
- whisk
- insulated cooler
- dish towels

Ingredients

- ½ gallon milk
- 4 tablespoons yogurt with live bacterial cultures

Directions

1. Heat the milk slowly until it reaches 180 degrees Fahrenheit, then remove the milk from the heat and let it cool to 120 degrees Fahrenheit.
2. Whisk in the yogurt.
3. Pour the yogurt into the quart jars, then put the lids on the jars.
4. Pack the jars into a cooler, keeping the yogurt at 100 degrees Fahrenheit for about 6 hours to inoculate the milk.
5. Remove from the cooler and chill in the fridge. While the yogurt is done, I like to let it stiffen in the fridge before consuming. Serve with honey and granola. The yogurt will keep for about 2 weeks in the fridge. That's really it.

Good bacteria

The bacteria cultures in yogurt are primarily *Lactobacillus delbrueckii* subspecies *bulgaricus*, *Streptococcus thermophilus*, and quite a few others to be honest. The bacteria consume the sugars in milk and emits lactic acid as a by-product, which accounts for yogurt's sour taste and stiff texture. Most of the bacteria used to make yogurt is believed to naturally occur in plants, strains of *Lactobacillus*, in the form of a liquid inoculant, have been used in the Yucatan Peninsula to combat mold infections on crops, which may account for yogurt's longevity.

Marble Method for Making Butter

If you want to try making butter at home, you can use this popular jar and marble method.

Equipment

1 clean quart-size mason jar with lid and ring

1 small, clean marble

Ingredients

2 cups heavy whipping cream

Directions

1. Put the chilled milk into the jar with the marble.
2. Shake the jar until butter forms, about 20–60 minutes.
3. Separate the butter from the buttermilk with a mesh strainer.
4. 4. Rinse the butter with cold water, then wrap it in wax paper and store in the fridge until stiff, approximately 12 hours.

CHAPTER 17: FERMENTATION

My first foray into fermentation

My first experience with fermentation came from Sandor Katz, the author of *Wild Fermentation* and *The Art of Fermentation*, who was the keynote speaker for the Sustainable Agriculture Conference put together by the Carolina Farm Stewardship Association every year.

I attended Katz's workshop in an effort to interview him about his work and as part of *From Scratch* magazine's coverage of the conference. I had a passing interest in fermentation at the time. I mean, who doesn't like home-brewed beer and wine?

During the class, Katz discussed the varied and rich history of fermentation. According to Katz, fermentation is just as much a part of human culture as agriculture, if not more so. It is a method of processing food to make it edible, a method of food storage, and a method for personal health care.

I sat enraptured as Katz discussed the beautiful history and fascinating science behind the alchemy that is fermentation. By the time it was over, Katz was my guru, I was his disciple, and *The Art of Fermentation* was my bible.

I interviewed him after the class, and he revealed his personal history and how fermentation led him to better health and a more nuanced spirituality. I went home and immediately started making sauerkraut the Sandor Katz way. I chopped up three heads of cabbage, adding salt to taste, massaged it, packed it into mason jars, and waited.

Three days later, after diligently "burping" it every morning, I cracked open a jar and tasted my magical creation. It was awful.

During the process of salting the kraut, by salting it "to taste," I had overdone it. Instead of a pungent, slightly salty, delicious treat, the sauerkraut tasted like a salt lick. Of course, I did not blame Guru Katz, as his methods were merely the wine poured into an imperfect vessel.

I went back to *The Art of Fermentation* and decided to actually follow a recipe this time. Instead of salting it to taste, I found it was best to use about two tablespoons of salt for every three heads of cabbage (which is still a little too salty for my wife).

The second batch was divine. I ate it for every meal: eggs and sauerkraut for breakfast, turkey and sauerkraut sandwiches for lunch; you get the idea. I even used it as a palate cleanser between dinner and dessert.

I still process as many fermented foods as possible, but the sheer longing my body has for these powerhouses of microbial activity means I hardly ever have

Fermentation takes cooking and turns it into mad science, or magic.

any in the house. But, that one-hour course has led to an obsession, and I don't expect it will let up any time soon.

What is fermentation?

Fermentation, simply put, is the process wherein microbes—yeast and bacteria, mostly—turn the carbohydrates in all sorts of food into acid or alcohol. The process also produces gases, like carbon dioxide and methane, but most of these are considered by-products and are not usually used, but these gases often produce desirable outcomes. In bread, the carbon dioxide produced by yeast makes the end result soft and fluffy.

In the case of wine, beer, and saké, yeasts turn the carbs into alcohol. In the case of acids, bacteria take foods like cabbage and turn it into kraut. Other common foods that are fermented are vinegar, kimchi, salami, chocolate, vanilla, bread, and yogurt.

A bit about botulism

Botulism is one of the biggest fears most people have when it comes to fermentation. It's mostly unfounded. Botulism has to have an ultra-low oxygen environment in order to grow and thrive. Because items like sauerkraut, kimchi, and other lactobacillic fermentations are exposed to oxygen, it is nearly impossible for the botulism bacteria to grow in these products.

Essentially, the bacteria and fungi responsible for fermentation are introduced into a food (think yeast into bread dough) and given an environment that encourages them to grow and flourish. As part of that expansion of microbial populations, the helpful little germs consume a portion of the food, breaking down molecules of proteins, fats, and carbohydrates.

The fermentation process goes back to the Neolithic period, which puts fermentation in contention for the earliest food technology created by humans. In fact, fermentation was so important to humans, there is a well-supported belief among some archeologists and anthropologists that agriculture—responsible for the beginning of human civilization—was spurred by early human's desire to produce beer, a theory backed by a paper published in 2012 by researchers at the Simon Fraser University in Canada.

The process allows those who consume the food to get more nutrition and caloric content; and in the case of beer, wine, mead, and other spirits, produces satisfying side effects.

Additionally, the consumption of fermented food helps supply the human digestive system with beneficial bacteria, which can help shield the system from disease, parasites, and other pathogens. In fact, Katz, who was diagnosed with HIV in the 1980s, told us during his interview he credits his continued good health, in part, to his high levels of consumption of fermented foods.

Fermentation also provides an excellent method of long-term storage. When talking about cheese (another fermented product) my father used to say, "Cheese doesn't go bad, it just gets better."

Fermentation essentially uses good microbes to prevent bad microbial growth, which is what causes food to spoil. As a result, fermentation was one of the first food storage methods.

While I love home canning, I love home fermenting even more. By fermenting food you get the same results as canning—food stored for a long period of time—without the trouble and equipment needed for canning.

Sauerkraut and beyond

If you're interested in fermenting, you should be aware that you can ferment nearly everything. I mean everything. I've fermented everything I could conceive of: cabbage, kimchi, Meyer lemons, limes, green beans, sourdough starter, green tomatoes, cheese, yogurt, basically anything you can eat; I even made vinegar from fruit scraps. I haven't done eggs or meat yet, as both usually require a bit more prep work, but there's a recipe for one hundred-year-old eggs I'm dying to try.

Is it necessary to inoculate?

In most cases of fermentation, you have to inoculate your food (see Chapter 16 regarding yogurt). In the case of sauerkraut and most lactobacillic fermentation, you don't have to inoculate your food, as the bacteria that ferments these products naturally exist on almost everything.

Fermentation takes cooking and turns it into mad science, or magic.

In a way, you're taking fundamental forces of nature, microbial metabolism, and using it to prepare food. It feels like cooking with the Universe as your stove.

For the most part I stick to bacterial fermentation. Other types of fermentation—cheese making, yogurt making, sourdough—are discussed in other chapters. As far as alcohol fermentation, this tends to require more time that I normally have. I've fermented beer and wine, but so far producing a consistent product is beyond my ability. I've included a great book on alcohol fermentation in the list at the end of this chapter if you're interested.

The easiest way for most Americans to begin fermentation is to start with sauerkraut. The ingredients are cheap and readily available, and most of the equipment can be found in the average home. It's also a great introduction to the foibles and fears that most people have about lactobacillic fermentation, including mold formation.

Troubleshooting

The biggest issues most people have with lactobacillic fermentation are mold and saltiness.

When I first started out, I used way too much salt. Fermentation tends to magnify the salt taste of food. So, if you happen to accidently put too much salt in, you'll definitely notice when you taste your kraut. If the salt content is low enough, you can rinse the kraut and reduce the salt taste. If it's too high, your best bet is to toss it in the compost.

Chlorinated water and iodized salt

Most fermenters and recipes call for unchlorinated water and non-iodized salt (sea salt, kosher salt, pink salt, etc.). The idea behind both is to make sure you don't retard the growth of bacteria in the ferment. Chlorine and iodine both kill bacteria. The iodine in salt, however, probably isn't a strong enough concentration to kill bacteria. As far as chlorinated water, boiling it for a minute or two should be plenty to remove the chlorine in it.

I've used both iodized salt and chlorinated water before, when I didn't know any better. The kraut I made turned out fine . . . that time. Other times, not so much. If you do use chlorinated water and iodized salt, your ferment will probably be fine, but since you want to maximize the chances of bacterial growth in your ferment, you should avoid chlorine and iodine.

Mold tends to frighten most people. As long as the mold isn't black, you don't need to be concerned. Blue or white mold is generally safe (barring a compromised immune system or allergy) for human consumption. So, if you see it on top of your kraut, scoop it off and add more brine. If you see black mold, toss the kraut and start over.

If you have trouble with too much mold, you may have detergent residue inside your crock. Dish detergent kills bacteria, but since you are using bacteria to facilitate fermentation, detergent residue inside the container can stunt the growth of the bacteria, which could lead to the establishment of mold colonies inside your ferment. To avoid this, make sure you rinse your crocks and bowls well.

Resources

If you'd like to learn more about fermentation, check out these books:

The Art of Fermentation and *Wild Fermentation*, Sandor Katz

The Mini Farming Guide to Fermenting, Brett Markham

How to Brew: Everything You Need to Know to Brew Beer Right the First Time, John J. Palmer

The Joy of Home Wine Making, Terry A. Garey

Sauerkraut

Equipment:

clean mason jars with rings and lids
large crock or bucket (no metal)
large towel
1-gallon ziplock bag
glass or plastic bowl

Ingredients

3 medium heads of cabbage
2 tablespoons sea salt

Directions

1. First, cut the cabbages into quarters, then cut the core out of the base of the cabbage. Next, chop the cabbage into strips about ½ inch wide. Some kraut makers use a grater for this process, but I like my kraut to have a big "crunch" to it. Put the chopped cabbage into a glass or plastic bowl. Never use metal, as the metal can react to bacteria in the cabbage, killing it.
2. Add the salt to the cabbage. Some kraut makers layer the salt as they put in the cabbage. I just dump it in and start stirring. Use your hands (clean hands!) to combine the salt into the cabbage. Then start squeezing the cabbage; just grab big fistfuls and squeeze. This mechanically pulls moisture from the cabbage leaves. The salt chemically pulls the moisture from

the cabbage leaves through osmosis. The longer you squeeze the leaves, the more moisture you'll pull out. You are finished "juicing" your cabbage when you squeeze a handful and it produces about as much moisture as a wet sponge.

3. Put the cabbage in the crock or bucket (again, ceramic, plastic, or glass—no metal!). If the cabbage juice does not cover the cabbage leaves in the crock, make a brine (2 tablespoons of salt to 1 quart of non-chlorinated water) and put in enough to cover the leaves.
4. Open the ziplock bag and place it in the crock. Fill the bag with water until the weight of the water pushes the cabbage leaves deeper into the crock and increases the water level. You want the weight of the water to push the brine about 1 inch past the cabbage leaves. Close the bag tightly and let it sit.
5. Cover the entire crock with a large cloth or towel, tying it off. Then put the crock in a somewhat cool location out of direct sunlight. In about three days, you should have a pretty good kraut. Taste the kraut and see if it's sour enough for your liking. If not, recover the kraut and let it sit longer. How long depends on you. Most people tend to ferment for a few weeks, sometimes a month. My personal record for fermenting on the counter is eight weeks.
6. Once it's fermented to your taste, take it out of the crock and pack it into mason jars. Cap it off with the lids and rings (plastic lids that fit mason jars work well for this sort of project, as kraut doesn't have to be sealed). Store the jars of kraut in the fridge. This doesn't stop the fermentation process, but it slows it down greatly. Eat it as often as you can!

CHAPTER 18: EGGS

Love at first crack

Chickens are supposedly the gateway drug to homesteading, but I would argue it is the fresh eggs that really get people hooked.

Fresh-from-your-chicken eggs, for me, were a revelation. Some of my fellow homesteaders can trace their beginnings to a CSA box that introduced them to fresh, local produce as the beginning point. Others point out the first time they cooked with herbs they grew themselves. For me, I knew I was on the right track when I had that first brown, dirty looking egg.

I had to rinse it in the sink, as it had a bit of chicken poop and straw on it. It felt different immediately. Chalky and rougher than store-bought eggs.

I cracked it against the side of the pan, like I always do, but it did not break like eggs normally do. Because the shell was thicker, it required a bit more force than I was used to, so I got shells in the pan.

The egg came out visibly slower than I was used to. The white was thicker, the yolk bigger.

I looked at that egg in the pan and realized how sickly my eggs looked before. The store-bought eggs were watery in comparison. The yolks in these eggs, eggs from chickens I fed every day, were deeper in color, more orange than yellow.

I scrambled them, and that bright, beautiful bit of color suffused the mix. I was so fascinated by the process and the outcome, I actually made six more eggs that morning. I fried two—one lightly and one well done—I boiled two and scrambled two more.

I ate them all.

The color and the texture and the flavor were incredible. That was about the time I knew my wife and I had the right idea about homesteading.

Eggs are great for homesteaders. They're an excellent protein source; they provide necessary B vitamins, which are good for vegetarians; and they have a nice concentration of calories. A large egg has eighty calories. Which is more than most homestead products. And since chickens are omnivores, they can consume things you might not want to, like earthworms, insects, vegetable scraps, meat scraps, etc., and turn them into amazing, delicious balls of nutrition. They come with their own storage system and keep a lot longer than you might think they would.

Egg surprises

When it comes to home-harvested eggs, there will be several surprises. You must remember, your store-bought eggs are sold with all the quirky eggs weeded out. So you might see some eggs that are nonconformist in presentation. While we're discussing chicken eggs here, it should be noted that this information applies to just about every type of egg you might produce on your homestead.

Fart eggs—Sometimes your chickens will produce miniature eggs. The mini eggs are called fart eggs. This may happen with birds that have just started laying eggs. After a chicken lays eggs for a while, these miniature little missteps in egg production will disappear.

Odd eggs—All over the country, the appearance of odd eggs, in dumbbell shapes, heart shapes, and all manner of size, used to be cause for contacting the media. During the pre- and post-World War II era, newspaper cub reporters were often given the task of photographing and writing up stories about the strangest eggs farmers brought into their offices. In rural areas of the country, it happened regularly enough that some reporters were jokingly given the title of "Odd Egg Editor." That practice has fallen by the wayside, but farmers, backyard chicken keepers, and homesteaders still get the occasional odd egg. Sometimes these eggs are a sign of sickness or physical damage, but most of the time they're just weird little things that happen. If you do get an odd egg, post it on social media to amaze all your friends, but don't panic. Just keep an eye on your feathered friends for a while and look for signs of sickness: lethargy, cloudy eyes, strange behavior, etc.

Bloody eggs—Sometimes there will be blood in your eggs. Several superstitions have surrounded the appearance of blood in the eggs. Some believed it was an omen for the death of a loved one, or there is the more benign belief that the chicken laying the egg was frightened.

However, there's a simple explanation: blood in an egg is usually caused by a blood vessel rupturing during egg formation. Think of it as a minor bruise. While startling in appearance, it does not make the egg inedible. If you want, eat the egg. If it is a bit too visceral (and it always was for me), just toss it.

Double yolks—Sometimes eggs have a double yolk. While this happens in store-bought eggs, I've noticed this is actually more common in home-harvested eggs. This usually means your chickens are well-fed and healthy. It also means you got lucky and get a little extra yolky goodness to start your day.

Embryos in eggs—Rarely, and only if you have roosters, you will crack an egg and find a bloody spot with veins running

Floating eggs

Want to know if your egg is bad? Float it. Put the egg in a bowl of cold water. If it sinks, it's good. If it floats, it's bad. If the egg stands on one end at the bottom of the bowl, it's a good idea to eat it that day, as it's on the verge of going bad. Let your kids check the eggs, they'll love it, and it's a bit of a science lesson for them!

off it. As startling as this is, it's actually a fertilized embryo. Most eggs are not fertilized, which is why you do not need a rooster to produce eggs. But, occasionally, a rooster and a hen will do what they have been designed to do, and a fertilized egg is the end result. Again, just like the appearance of blood in the eggs, this is harmless and actually edible, but feel free to toss it if it's just too much; you won't be the first person to do this.

Bad eggs—Chickens like to hide eggs. In fact, you will often have one or two birds that seem to hate the idea of you getting their eggs. In my opinion, this is usually the sign of a particularly bright chicken, as it is natural behavior from birds attempting to avoid egg-stealing predators in the wild. Technically, you're also an egg-stealing predator, so it makes sense that a hen might try to avoid your prying hands. Unfortunately, sometimes this means you will not find a clutch of eggs until they go bad. And doubly unfortunately, you will sometimes accidently crack one of those eggs into the pan one morning. It will smell like death warmed over. Just toss it out, there's not much that can be done. Clean the pan, burn some incense, and meditate in order to forget the memory of that awful, acrid smell.

Safety information

Fresh eggs can be kept for about two weeks without refrigeration, as long as they are not washed. Cleaning eggs removes a coating that prevents oxygen from getting inside the egg, and oxygen is what makes eggs go bad. Washed eggs must be refrigerated or used immediately after washing in order to prevent spoilage. Eggs should be kept at 45 degrees Fahrenheit or colder after washing. If kept at the proper temperature, eggs can be stored in the refrigerator for up to five weeks. Eggs can be frozen, but not in the shell. I freeze eggs in ziplock freezer bags. If you do freeze the eggs, whisk them lightly beforehand. According to the Egg Safety Center, eggs can be kept frozen for a year.

Goose eggs are noticeably larger than duck and chicken eggs. Getting the eggs from ducks and geese may prove to be a little more challenging than getting them from chickens.

If you have not washed your eggs before storing them, be sure to do so before using them. Fresh eggs often have poop and bacteria on them from inside the chicken coop. You can give them a soft scrubbing with a dish sponge under warm water. The whole plan is to prevent fecal contamination while cooking.

Tips for getting eggs from other birds

While chicken eggs are the most common on homesteads, other eggs are also produced by homesteaders. Ducks, geese, and quail all produce eggs that are suitable for consumption or sales. Duck eggs are actually eclipsing chicken eggs in popularity, especially for restaurants and Asian food markets. All these eggs are handled almost identically after laying, but here are a few tips and tricks for helping your birds produce eggs.

Ducks and geese

Ducks and geese tend to handle their eggs strangely. In nature, they constantly hide their nests, and since they aren't quite as domesticated as chickens, they either find the oddest places to leave their eggs, or just drop them in the middle of the run.

You can build and set up nesting boxes for them, which will encourage them to keep their eggs a bit better for their human partners. The easiest way is to set up straw bales. First, set up a single row of bales, end to end, to serve as the backdrop for the boxes. Then, place bales perpendicular, just far enough apart for one bird to sit inside. Then, put at least six inches of straw in each nook. Geese and ducks like to "bury" their eggs in debris, so by providing it you're encouraging them to use the nesting spots you created.

You might want to mount a one-by-six piece of untreated lumber along the front of the nesting spots. Drive a few stakes into the ground, then screw the board to the stakes. The ledge created will help keep the straw inside the nest areas.

In my experience, I have not noticed that one particular type of straw is favored over another. I've used pine straw bales and wheat grass bales and had success with both.

As you remove the eggs, make sure you replenish the straw as necessary. If your birds can't bury their eggs, they will abandon the nest areas and seek out better hiding spots.

If the nesting area is outdoors, it will need some sort of structure to protect the inside from the elements. Geese and ducks are not really susceptible to cold weather, but it is best to ensure they're as comfortable as possible. While much larger, geese and duck eggs are nutritionally similar to chicken eggs by volume.

It takes about five quail eggs to equal the volume of one chicken egg. Despite their small size, quail eggs are popular at restaurants and specialty markets.

Quail

Quail are often kept in small hutches, although some keepers advocate keeping them in the same conditions as chickens. I lean toward the camp that advocates coops and runs, but quail are much better flyers, and as such they often require much tighter fencing to keep them in captivity. If they're kept in coops, build them boxes similar to chicken nesting boxes. However, make sure they have plenty of straw. Quail like to build nests and will use a lot of straw to do it.

If kept in a hutch, provide the birds with enough straw to keep them comfortable. Hutches placed on stilts will provide the birds with a bit more security from predators, and wire bottoms will make it easier to clean them.

Most quail hutches are made with slightly slanted bottoms, so the eggs they lay will roll toward the front of the hutch for easier collecting.

Quail lay more eggs than chickens or ducks, usually about one a day, but the eggs are much smaller. It usually takes about five quail eggs to equal the volume of one chicken egg.

Despite this, quail eggs are popular at restaurants and specialty markets. The eggs are also considered healthier substitutes for chicken eggs, especially for people with auto-immune issues.

Egg sales

If you decide to sell your eggs, the first thing you should be aware of is you cannot compete in price with store-bought eggs. Chicken eggs are one of the few agricultural products that benefit from economy of scale, which means the more eggs produced, the cheaper they are to produce. Most egg sellers seriously undervalue their eggs, often selling them for two dollars per dozen.

Larger homestead operations may use them as a loss leader, drawing in customers who want eggs, and then buy other items that have higher profit margins. If you decide to adopt this strategy, make sure you do the math and check it twice. Egg sellers who lose money will not be able to sustain sales levels. I would suggest that, at minimum, chicken eggs be sold for four dollars per dozen, more for organically produced eggs. That's approximately the break-even price for eggs laid by chickens fed conventional feed in the Southeast. However, that number might not even be a break-even point in locations where chicken feed is more than fourteen dollars per bag.

You should also charge more for specialty eggs: goose, duck, and quail eggs are often harder to find, and demand drives price just as much as cost of production.

Be aware, egg sales are usually highly regulated at farmers' markets. Agriculture and health officials will check storage temperatures, inspect washing stations, and make sure egg cartons are new. While these regulations are not onerous, it's a good idea to find out what your state requires for egg sales to avoid any fines.

To avoid the cost of purchasing cardboard egg cartons (don't buy Styrofoam) you might want to consider selling your eggs in plastic, reusable egg cartons. If you do, you can offer a discount to customers who bring their cartons back to reduce your cost and reduce landfill waste among your customers. They're sure to appreciate it.

Washed versus unwashed eggs

American eggs cannot be exported to most European countries, because American eggs are washed, and Europe won't allow them. Conversely, the eggs produced in Europe wouldn't be allowed on American shores, as the unwashed eggs wouldn't pass inspections here.

If you're not selling your eggs, it's up to you whether you wash them or not. The natural coating on eggs, the "bloom" as it's called, prevents the egg from going bad for about two weeks without refrigeration. If you eat your eggs fast enough, you can probably keep them stored on the counter. Again, though, if what you're about to eat seems "off" or bad, don't eat it. It's not worth the risk.

CHAPTER 19: THE BASICS OF BAKING

How baking bread saved my sanity

Before I started my homesteading journey, like many Americans after the financial collapse of 2008, I was laid off from my job. It was not my fault; I was just downsized.

Many people all over the country faced the same situation: the mortgage bonds collapsed, banks were bailed out, and the little people of the world got laid off.

It was a rough year, to say the least.

Immediately after being laid off, I did everything a person is supposed to do when they get laid off: I worked desperately hard to find another job while trying my best to feel useful. I kept the house as best I could, I packed lunch for my wife, who luckily managed to hang onto her job, and I took care of the kids.

One day, while hanging onto my sanity by cleaning the kitchen, I found a bread machine my wife had purchased at a thrift store during one of her treasure hunting trips.

Suddenly, I had a plan. I'd make bread! That would save my family money and give me something useful to do.

The bread machine saved my sanity that year. I searched online and downloaded a pdf of the owner's manual—bread machines are not intuitive appliances. I only used it about half a dozen times. Despite being easy to use, making bread is an art, and bread makers don't often allow for the expression of the baker, or the yeast for that matter.

But it introduced me to the world of baking, and I'll be forever grateful to that little five-dollar thrift store find (and my wife, who presciently purchased it).

I spent the next year—until I finally got a job at a media company—baking bread for my family. My kids loved it, my wife loved it, even my in-laws, who stayed with us for a brief period during that time, loved it.

I baked every kind of bread I could think of: pita bread, sourdough bread, wheat bread, Amish sandwich bread, French loaves, you name it, I baked it.

Looking back, that was probably the year we started thinking about homesteading as a lifestyle.

Baking bread showed me and my family that maybe, just maybe, we didn't have to buy into the consumerist culture. I could bake bread for myself and my family. Maybe I could do more.

The science behind the magic

Baking, like fermentation and agriculture in general, is an integral part of human history. Bread is mentioned in the Bible, depicted on the walls of Egyptian temples. Goddesses of baking and bread have been worshiped for millennia.

Bread allows humans to process cereal grains—wheat, barley, rye, oats, etc.— into consumable food. At the same time, the process of adding yeast to dough and letting it rise, allows people to get more out of their grains. Bread is primarily composed of a handful of ingredients: flour, oil, salt, and water. The outcome, however, is greater than the sum of its parts. Bread contains proteins not found in flour, for example.

While bread and the gluten that binds it together have gotten a bit of a bad rap recently, I'm personally of the opinion that has more to do with the way wheat is produced in our food system, as well as the industrial methods of food preparation.

The process of producing bread is composed of a handful of steps, no matter the recipe: proofing yeast or sourdough, mixing dough, kneading, first rise, kneading and forming dough into loaves, second rise, and baking.

The proofing and rising parts of the process are, in my opinion, when the magic actually happens.

Proofing is the activation and reawakening of leavening agents. For almost all bread, the leavening agents are yeast or sourdough starter. Yeast is composed of just that, living yeast microbes. Yeast is a fungus that consumes sugars and carbohydrates, producing alcohol and carbon dioxide.

The small amount of alcohol produced by bread yeast is cooked off, but the carbon dioxide provides gas bubbles throughout the dough that give most bread its fluffy texture and causes the bread to rise. As the gas is produced, the dough expands to accommodate the bubbles.

In the case of sourdough, the starter is actually a SCOBY (symbiotic colony of bacteria and yeast). Like pure yeast, the wild yeast in sourdough produces carbon dioxide, but the bacteria in the starter produces acid, which gives sourdough its distinctive flavor.

What is gluten?

Gluten is a binding protein produced during the bread making process. As yeast consumes carbohydrates in the bread, proteins in the flour are left behind. The proteins, called gluten, help provide bread with the structure it needs to contain the carbon dioxide produced by the leavening. White flours have a higher gluten concentration than whole wheat flours. In fact, many whole wheat bread recipes call for extra gluten, or a combination of wheat and white flours to produce the necessary "softness" most people look for in bread. A pure whole wheat bread tends to be more dense than bread with extra gluten or with white flour as an additive.

Bread and beyond

Once you get bread down, you will want to move onto other projects. Muffins and cakes are pretty much just sweetened bread. And, since most muffins and cakes call for chemical leavening agents, they don't require the rising time that bread does. Pie crusts, despite their reputation, are actually easy to make. I highly recommend using lard instead of shortening, as it will result in a flakier crust. Once you have pie crusts down, not only can you make your own fruit pies, but you can also experiment with meat pies. This will make you the most popular person in your family, and probably in your neighborhood.

Leavening agents

While most bread is leavened with yeast or sourdough, many baked goods call for self-rising flour. Self-rising flour has baking soda and baking powder mixed in. Both of these ingredients create a rise in breads and other baked goods by producing carbon dioxide through a chemical reaction. In fact, some recipes even call for a tablespoon of vinegar to help make cakes and muffins fluffier.

A similar process is produced in a Southern-style cake, called a Cola Chocolate Cake. Instead of using baking soda or baking powder to produce carbon dioxide, the recipe calls for 8 ounces of cola. The carbonation in the soda actually provides the gas needed for the cake to rise.

Baking tips

If you get deeply into baking, there are three things you can do that improve your chances of success.

1. Use bread flour—You can avoid it, substituting all-purpose when a recipe calls for bread flour, but bread flour makes a big difference in the texture of your bread. Bread flour has more protein, which leads to more gluten structures.

2. Warm up the oven—Before you mix your dough, turn your oven onto the lowest setting. As soon as it's done preheating, turn it off, but don't open the oven door. Once you're done mixing the dough, set it inside the oven to rise. The extra warmth will help promote yeast growth. Due to climate control, most modern homes have kitchens that are just a bit too chilly for yeast to do its thing. Giving it a warm place to incubate means you'll get a better rise from your dough.

3. Let sourdough sit overnight—Sourdough starter makes bread taste acidic. To get the best flavor in your sourdough, let it sit overnight. You'll get a better rise, better gluten formation, and a tastier bread if you let it proof for at least twelve hours, more if you want. The great thing about sourdough starter is you can play with it a bit, experimenting with different proofing times to get a different flavor and texture.

Pita Bread

My favorite bread recipe out of all the breads I've made (and failed to make) is a simple pita bread recipe. This recipe produces neat, fluffy little pita bread loaves. Children love them, and they make great sandwich pockets.

Equipment

baking sheet, preferably ceramic
metal tongs

Ingredients

1 cup warm water
2¾ cups all-purpose flour
2¼ teaspoons (or one package) yeast
1½ tablespoons olive oil
2 tablespoons salt

Instructions

1. Put the water, about 1 cup of flour, and the yeast into a bowl and let it proof for about 20 minutes. The mixture will be foamy and smell yeasty when it's ready.
2. Add the rest of the flour, oil, and salt, then stir until a sticky ball of dough is formed. Knead the dough in a stand mixer or by hand for 5–8 minutes, until it feels springy in your hands.
3. Oil a bowl large enough for at least twice the amount of dough you have, then place the dough inside. Turn the dough, coating it lightly all over with the oil. Cover the bowl with wax paper and a towel. Let the dough rise for about 2 hours.
4. Once the dough has risen, cover your work surface with a light dusting of flour. Knead the dough again for about 2 minutes. Break the dough into eight roughly equal size pieces, making balls with the dough. Place the dough on the work surface, sprinkle it with flour, and cover it with a towel. Let it rise again for about 30 minutes.
5. While you're waiting for the dough to rise, oil the baking sheet. Turn on the oven to 475 degrees Fahrenheit and put the baking sheet inside so it can warm up.
6. Once the dough has risen, take each ball and flatten them into circles, about 8 inches wide and ⅛ inch thick.
7. Put one of the circles of dough onto the baking sheet. Let it bake for about 2 minutes, then flip the loaf over and cook for 1 minute on the other side. Remove the pan from the oven and place the bread in a bowl. Repeat step 7 for the rest of the dough circles.

Sourdough Starter

If you want to make bread, you'll eventually want to make sourdough, and for that you'll need a starter. If you have friends who bake, you can ask them for some of their sourdough starter—this is my favorite way to get a starter, since you learn more about baking from others. However, if no one you know bakes, then you can make your own.

You should think of your starter as more of a strange pet than a baking ingredient. If properly cared for and fed, the starter will last decades. The older the starter gets, the more effective it will be.

Equipment

quart-size mason jar
dish towel
elastic band

Ingredients

whole wheat flour
non-chlorinated water

Instructions

1. To start, combine 1 cup of flour and ½ cup of water in the mason jar and mix it together completely. There shouldn't

be any dry spots at all when you're done.

2. Cover the jar with the dish towel and use the elastic band to hold the dish towel to the lip of the jar. Don't let the starter come in contact with any metal, as metal can cause a reaction and stunt the growth of bacteria in the starter. Store the starter mixture away from direct sunlight on your kitchen counter, or I find it does better on top of my refrigerator, as the heat in the kitchen rises, encouraging growth.
3. Twenty-four hours later, throw half of the starter mix into your compost (or give it to a friend) and replace it with ½ cup of flour and 1 cup of water, more if the starter gets too dry. For reference, it should be the consistency of slightly thick flour paste.
4. Repeat Step 3 for six days, feeding it daily. Through this period, the starter will bubble, foam, and produce a sour, yeasty smell. It may also produce a clear, alcohol-scented liquid on top. This is called beer. If it does produce a beer, discard the beer and feed the starter as normal.
5. At the end of six days, you should have about 12 ounces of starter. If you don't, add more water and flour to make up the difference and continue the process for three more days.
6. Store the starter in the door of your refrigerator with the dish towel and elastic band intact. Once a week, feed the starter as previously explained. Whenever you make a recipe using your starter, keep at least 4 ounces of starter in the jar, then add fresh flour and water to make up the difference.

Mom's Cornbread

While not as glamorous as bread made from wheat flour, cornbread is a staple food in the South. It takes almost no time to produce, and it gives families a nice bit of starch to go with a spicy bowl of chili or a big pot of black-eyed peas. It was also one of the things my mother excelled at cooking.

Here's her recipe. My mother didn't really believe in measurements, so you may have to play with the amount of the individual ingredients to get your cornbread right.

Ingredients

- 2 cups cornmeal mix
- 1 egg
- 1¼ cups buttermilk
- 4 tablespoons mayonnaise

Instructions

1. Preheat the oven to 400 degrees Fahrenheit.
2. Combine all ingredients in a bowl and whisk them together until they are well incorporated. This recipe uses mayonnaise instead of oil, which makes for much moister cornbread. While the recipe calls for 4 tablespoons, my mother (and I) simply takes a tablespoon and scoops two healthy globs of mayo right into the mix.
3. Pour the mix, which should resemble thick batter, into a greased cast iron skillet. If you have it, lard is the best thing to grease the skillet with. If not, you can also use oil or cooking spray.
4. Bake for about 20 minutes. When the cornbread is ready, the edges will separate from the skillet, and a fork poked into the top of the cornbread will come out clean.

6 | Section Six: Lifestyle

$E=mc^2$

CHAPTER 20: CHILDREN

Homeschooling

When our son was in elementary school, he attended public school in South Carolina. Having been diagnosed with dyslexia when he was five, we'd actually been through several schools in three states, including a charter school, in order to find him a suitable education system to deal with his unique learning style with very limited success.

Finally, a tutor that was hired by the school system specifically to deal with his needs, said the best thing would be for us to homeschool our son. It scared all of us—me, my wife, and my son—to back away from the system and program we'd known all our lives. At the time, the homeschooling movement was doing well, but it wasn't the ubiquitous concept it is today.

So, we did some research, contacted the state agency that handled homeschooling, and put together a curriculum tailored to our son's needs.

We even got a desk, a whole bunch of school supplies, and a whiteboard to teach him lessons on.

The shopping was fun, but we quickly realized that the goal of homeschooling wasn't to replicate a school in our home, but to use our home as a school.

With a few hours of education per day, we managed to get our son caught up to his grade level and found strategies to help him deal with the dyslexia. We came to view it not as a disability, but as a new way of learning, specific to him and children like him.

While we occasionally worked with homeschooling groups in the area, we wound up taking a student-focused approach that took in all that we experienced and dealt with in our house. He learned fractions while cooking. He counted money during shopping trips, he learned about life sciences on hikes and trips to state parks. We visited museums and aquariums—anywhere there would be a learning opportunity. He used comic books to expand his vocabulary and get comfortable with reading.

He graduated high school in 2016 and started college in January 2017.

When we started homeschooling, we were worried he'd be socially stunted. However, during numerous outings, both with adults and children his age, as well as opportunities to participate in sports, library programs, 4-H clubs, and art classes, he managed to make all sorts of friends. Honestly, the socialization issue just wasn't that big a deal.

Lessons learned on the homestead are applicable to just about every area of study. My daughter started her own business at the age of ten and continues to thrive.

In fact, we enjoyed the process of his homeschooling so much that we decided to homeschool his baby sister as well. She's in the fifth grade and is thriving.

Homeschooling and homesteading go together like peanut butter and chocolate.

Having kids at home helps keep homesteaders grounded, provides a little extra labor, and the lessons learned on the homestead are applicable to just about every area of study. If you want your kids to learn about math, take them to the farmers' market and have them handle the cash while you sell vegetables. Teaching them writing and reading? Have them write a blog post for your social media site. Need to work on handwriting? They can make signs! Science is obvious, as a plethora of growing things, bacterial activities in the kitchen, and animals running all over the place provide a chance to talk about everything from climatology to biology. The changing seasons allow for a discussion about the solar system, and using the farmer's almanac as a guide to planting crops gives you the chance to discuss astronomy.

If you have more than a couple of children, you'll need to get organized. Kids require structure, so even if you don't set up your own little school corner, you still need to establish a set time during the day for lessons. Once your kids are older, they don't necessarily need as much one-on-one instruction, so you can provide them with lessons and exercises and be on hand to provide help or extra tutoring and explanation if needed. I've found the best way to provide this is to plop them at the kitchen table while you get some chores done. If you have multiple children, make them help each other. They can explain things to each other, call out words for spelling tests, and so much more.

Homeschooling also gives homesteading parents a fascinating opportunity to watch children grow up in an environment that barely exists anymore. Your kids can feed chickens, milk goats, pick strawberries, and eat apples straight from the tree. They can learn about where food comes from, how things are made, and so much more.

My son, for example, fell in love with our chickens when we first started homesteading, and those hens fell in love with him too. During the winter, he would sneak off to the chicken run, coax a chicken into his lap, and zip her into his jacket so they could keep each other warm.

Raising your kids on the homestead will give them a can-do attitude and lots of things to be proud of. When my daughter was ten, she started her own business selling plant starts in newspaper pots.

My daughter, at the age of ten, started her own business selling plant starts in origami newspaper pots at the farmers' market.

While both of them complain nearly constantly about the chores and endless amount of work, truthfully, a life filled with animals, baking, crafts, crochet, and harvesting tomatoes is a pretty good one for children, chores and all.

If you're interested in homeschooling, there are dozens of curriculums and resources for you, but the most important thing you will need is a bit of courage. Homeschooling tends to freak people out. There's a false belief that homeschooling parents and their children are abnormal, but studies have shown that they turn out just as well as their counterparts who attend public and private schools.

Homesteader profile: Jennifer Hall

Jennifer Hall has been homeschooling her children since 2000, right before her oldest would have entered kindergarten. Since then all of her children have been homeschooled. Her oldest, Ryan, the first to graduate from their homeschool, is studying physics and preparing to enter the Air Force.

Hall has eight children: Ryan, twenty, who holds an associate's degree, is in his junior year of college pursuing a bachelor's degree in physics. Noah is seventeen and in twelfth grade. He's also enrolled at the local community college. Hannah is sixteen and in tenth grade. Elijah is thirteen and in eighth grade. The family also includes Jonah, eight; Josiah, seven; Moriah, five; and Zarah is three.

"Home education made sense for our family for numerous reasons. When we began, my husband was active duty military, which we expected would entail relocating every few years, and homeschool would allow us to continue schooling without interruption or switching schools," Hall said. "We wanted to educate our children in line with our religious beliefs, and we felt that it was our responsibility as parents to decide when our children would be exposed to sensitive topics. We felt that we could provide a flexible and dynamic education based on and tailored to each one of our children's personal needs."

Hall said their school has evolved and changed over the years based on the needs of the children, where they lived, her husband's deployments, and having more children. She said they just figured out what worked for the family.

The family has participated in homeschooling co-ops, support groups, 4-H, vocational training, and have enrolled for college-level courses while still at the high school level of their homeschool.

"Over the years, we have been pretty eclectic in regards to curriculum choices. We prefer a mix of unit studies (mostly in the summer months), lots of reading, child-led lesson plans with a core of phonics, reading, grammar, and mathematics. All of the children love and devour books, we have weekly trips to the library to supplement our fairly extensive home library," Hall explained.

The biggest benefit, she said, was the time the children have to explore areas and topics they're interested in.

For example, one child has developed an extensive yoga practice, another plays the guitar, and another is building a pollinator garden for bees and butterflies.

The family is also able to spend more time outdoors, hosting classes, and just getting lots of sunshine and exercise, swimming and gardening and learning even through the summer.

"Our school operates on a year-round schedule January through March with a

Resources

To learn more about homeschooling, check out these books and websites.

Websites

Khan Academy—This website is devoted to providing lessons and exercises for everyone. While the site is light on the liberal arts, like English, its programs for math and science are unmatched. Find out more at www.khanacademy.org.

Duolingo—Free language learning software that provides instruction in twenty-seven languages, including English as a second language. This site has been compared favorably to other more expensive programs. Like Khan Academy, Duolingo is a nonprofit and exists solely to provide free language learning to everyone. Get started at www.duolingo.com.

Books

The Well-Trained Mind, Susan Wise Bauer

Home Learning Year by Year, Rebecca Rupp

One challenge on the homestead is getting kids to focus on their tasks, especially if you have animals.

break around Easter in April, May through July with a break in August usually around [the time] when public school resumes, and September through November with a break in December for Christmas fun and family time," Hall said. "For me, giving our children a strong foundation in the core subjects, allowing them to find what delights them, providing them with a lifelong love of learning, and building good character is the goal of our homeschool years."

Herding cats—raising kids on the homestead

Even if you decide not to homeschool your kids (no judgement) you will still notice some big differences in the way kids are on the homestead.

For starters, it's going to be hard to get them to focus on their tasks. There's going to be a lot going on. If you have animals, they'll want to feed (or play with, as the case may be) them. They'll get mesmerized by the oddest things—my kids get obsessed with spiders and lizards.

If you send them out to do chores, a task that would take you thirty minutes or so, the kids are liable to come back after a couple of hours with very little explanation of where they had been and what they were doing.

Annoying as it may be, this isn't a bad thing.

However, if you have multiple children, they'll form a bond that other kids raised in more conventional households probably won't have. All that time outside keeping an eye on each other while chasing lizards and daring each other to touch spiders tends to bind them together in a way that won't happen when children sit on opposite ends of a couch staring intently at screens.

It does mean that you will have to develop organizational skills you may not have needed before.

Your children are going to be a source of labor for you. To use that labor efficiently, you will have to make sure your kids know what's expected of them and when. For example, they need to know when animals should be fed—since unhappy animals are less productive. They'll need to know when to water the gardens and when to pick a harvest. Your kids will be responsible in ways that other children are not.

Chore charts are indispensable. Chalkboards and family calendars are going to be a big help. Make sure that seasonal items, like crop harvests and the births of baby animals, are accounted for. You do not want to be in a position where your three children have a soccer game and a karate tournament on the same day a crop of okra must be harvested.

Every time you add a project or an animal, everyone in your family will be impacted, and only by making your family a part of those tasks will you be successful.

Besides, it's good for them. By pitching in around the homestead, your children will learn things about responsibility and accountability, as well as real-world skills, like gardening and animal husbandry.

Raising your kids on the homestead will give them a lot of things to be proud of as well. Several times I've caught my children bragging to friends and family members about all the things they know how to do. And they've done it all, from injecting antibiotics into goats to planting and harvesting squash and corn to lancing and cleaning boils from injured and infected rabbits.

Helicopter parenting is out on the homestead; fortunately, it allows children the freedom to explore a little more, try things out, and gain confidence.

There's a bit of a can-do attitude that comes with homesteading. The more you do, the more you realize you can do. Your children are the same.

My kids and the children of my homesteading friends and neighbors tend to possess a confidence that many of their peers lack. For example, once my son decided to take his seven-year-old sister and a friend of his hiking in the woods behind our house. After about an hour he got lost and wound up on the backside of a nearby subdivision. Then they got caught in a rainstorm. After calming down his friend, he fashioned a makeshift umbrella out of a real estate sign for his sister, went to the nearest house, knocked on the door, and asked to use their phone. He called and asked us to pick him up, remaining completely calm the whole time.

The caveat to this confidence is you're going to have to avoid helicopter parenting. For starters, if your homestead is operating well, you will often find yourself in a position where you cannot stand and supervise your kids while they do a task. Which means you won't be able to hover around them while they administer worm treatment to the goats, since you need to weed the garden rows.

It also means they're going to be "free-ranging" more often than not. They'll spend time wandering the woods behind the house, catching frogs and fireflies, getting into poison ivy, and so very many more things.

In short, you're going to have to give your kids a little bit more room to screw up and have fun. Don't worry, it's good for them.

CHAPTER 21: COMMUNITY

Moving into a community

Immediately after moving to our new home in North Carolina, we knew we had a lot of work in front of us. Despite the property being home to apple, peach, and fig trees, as well as about one-eighth acre of blackberry bushes, outside of the grass being cut, the property had been untended for a while.

We had to mulch, establish raised beds, remove about half an acre of sod, and generally just get everything ready for all our animals and the crops we wanted to grow.

Our solution: we held a crop mob, potluck, and bonfire.

About half a dozen community members came, and we worked for about six hours. Some left early, some stayed late, but we made a lot of friends, and we got a great deal of work done.

Homesteading, if done properly and with compassion, builds community.

The importance of community

In order to homestead properly, establishing a community is key. I've been to many workshops and classes on homesteading and farming, and I've actually seen people break down in tears in the middle of spirited, positive discussions about the benefits of compost or the joys of seed saving. Many of these people were crying because they had been so isolated before the event.

Homesteaders are a quirky, and, quite frankly, weird lot.

Imagine the type of person it takes to enjoy activities most people have given up as frivolous or pointless in modern life: why grow your own veggies? They sell those at the store. Why have chickens in your backyard? Animals are stinky, and eggs can be had anywhere. Why make your own soap? That's just crazy.

Many homesteaders find themselves isolated, either geographically or ideologically. Their family and friends don't understand their desires to do what they do, and sometimes even having a discussion about the food system can lead to arguments and recriminations.

So as a homesteader it's important to reach out as often as possible.

Join mommy groups, homeschool co-ops, and take classes anywhere you can, like a master gardener's program at your local extension office. Volunteer at the farmers' market, at soup kitchens, or at 4-H outings.

Get out in the world and talk to other people who are interested in similar things. Isolation may be easier, but it won't improve your situation. And if you're going to homestead, then you need as much help as you can get. The people you meet through these activities are going to be your friends, customers, and

To avoid the isolation that many homesteaders feel, it's vital to build a community, whether in person or online.

helpmates. They will provide you with information such as where to get the cheapest grain for your chickens, the best deal on mulch, or where to get hay when you need it. They'll also be the people who show up when you need to pick six hundred pounds of zucchini or you have eggs to give away.

They will also help keep you sane and grounded. By providing a social outlet, the community you work to build will keep you from getting stuck in your own head and help prevent you from burning out.

Building a community

What if you don't have those types of things to connect with in your community? Believe it or not, some areas are so rural that the nearest library might be forty-five minutes away.

In that case, you're going to be responsible for building and nurturing your own community.

Have business cards printed up and give them to everyone. Collect phone

Reaching out online

Social media sites, like Facebook, Pinterest, and Reddit provide a chance for you to make connections easily and safely. If you're not comfortable having strangers come into your home, you can join a homesteading group on Facebook, or you can post on a community forum for your town or county and learn more about the people in your area. You don't have to welcome every stray person you find into your home, but you can still reach out to others in a safe, sane manner that takes your personal comfort level into account.

numbers and email addresses from anyone that expresses an interest in what you do.

Then, once a month, have a bonfire and invite everyone over or hold a potluck and share recipes. If you're really ambitious, you can put together a crop mob.

Ultimately, nurturing connections and relationships will only make life better for you and your community.

How to host a crop mob

Crop mobs were invented in Pittsboro, North Carolina, in 2008. Farming is labor intensive, organic farming even more so. Since modern families usually aren't as large as farming families of the past, it's necessary to find creative ways to get extra labor when you need it. So, a group of community members in Pittsboro decided to start a crop mob. Essentially taking the idea of a flash mob and applying it to farms.

Community members come together and work on a small, local farm. It's that simple.

But, it's also a bit more complicated. So, here's how to get your own crop mob started.

Step one: Figure out what you want done. If you need your beds weeded, your okra harvested, or your fruit trees mulched, then jot it all down. Whatever it is, have a plan for what you'd like a crop mob to do. There's nothing worse than showing up at a crop mob as a volunteer and the farm owner or homesteader doesn't have a plan for you.

Step two: Get your space ready. You want to have everything in place for your crop mob, so when the volunteers show up they're able to get right to work. If you want them to hoe your corn, then have hoes ready. If you need someone to spread mulch, make sure you have a shovel and a wheelbarrow (and the mulch). It's not about being a slave driver, it's about respecting others' time. If they have to wait for you to gather tools or find equipment, then they'll feel like they're being taken for granted.

Step three: Feed everyone. You don't have to put out a four-course spread, but your cop mobsters will be expending a lot of calories to help you out. It's only right and proper that you try and replace as many as possible. A pot of chili and a few loaves of bread with some cold lemonade will have everyone leaving happy and satisfied. In addition to feeding them, make sure everyone is as comfortable as they can be. Provide water, break locations with benches and shade, and access to the bathroom.

Step four: Relax and have fun. You may feel like you have a ton of work to get done, but remember, everyone who shows up is a volunteer. If a crop mob shows up, does an hour's worth of work, but then spends the next six hours chatting and socializing, don't get too upset about it. That's still an hour's worth of work you don't have to do. Crop mobs are social events that use work as an excuse to get people together. Don't get too hung up on the outcome.

Step five: Be grateful. Take the time after the crop mob to send out thank you notes. Get everyone's contact information before they leave. An email works, but if you can send out a good, old-fashioned note via mail, all the better. Additionally, you might want to consider a small gift for all your crop mobsters. Nothing showy or flashy, just something that lets them know you appreciate them showing up. A potted flower, a basket of veggies, or a plate of cookies lets everyone know they're appreciated. It also makes it more likely that they'll show up for the next crop mob.

FROM SCRATCH
8 HERBS
INDOOR
GET YO
COOP
GARDENER'S ONE LINE A DAY
A FIVE-YEAR BOOK OF GARDEN MEMORIES

CHAPTER 22: MARKING TIME

Living in the seasons

The best aspect of homesteading, as far as I'm concerned, is the mindful appreciation of the passage of time. Homesteading, no matter how it's practiced, involves being attuned to how the seasons move through your life. There's a season for everything you do on a homestead. Strawberries are bought when they're freshest. Canning food happens at the harvest. Winter means rest and planning.

Despite the fact that every aspect of homesteading is new every day, there's something eternal about living in the seasons and being acutely aware of them.

After all, spring, summer, fall, and winter come every year at the same time. Somehow, by plugging into those seasons, it makes me feel more a part of the world, like something about me, no matter how old I get or what happens to me, will come around every year too.

It's difficult to describe unless you get to experience it for yourself, but it's well worth experiencing. It's also something that most people in America don't get to experience anymore. Most people tend to spend the bulk of their time at work, which, truthfully, is a good and noble thing. It does mean, however, that

those same people are going to spend a lot of their time locked away from the elements. They'll go to their climate-controlled offices to work all day. When it's time to go home, they climb into their climate-controlled cars and drive to their climate-controlled homes.

Homesteaders don't. Homesteaders, especially if they also work away from home, are constantly in sync with the seasons. As a result, there's a bit less pressure, as homesteaders know that the flowers will bloom part of the year, the kale will be sweetest another time of the year, and the apples will ripen at yet another time.

Faith and homesteading

Homesteaders seem to overwhelmingly be people of faith. Even the agnostics and atheists among homesteaders tend to live in reverence of nature. It makes sense, actually. Culturally speaking, most religions, at least at one time or another, used their holidays and festivals to mark the changing of the seasons. Passover and Easter are both indicators of the beginning of spring. Pongal is a harvest festival celebrated by Hindus. Christmas marks the beginning of winter.

Those festivals and holidays, which are intertwined with culture and faith, are markers of significant agricultural times and phenomenon.

Celebrate the seasons

But, how do you mark time if you aren't religious? There's dozens of ways to do it. Astronomically speaking, you can mark the beginning of every season based on the position of the stars. The dates are provided by the National Weather Service, as well as the farmer's almanacs.

Whenever those times of the year come around, take some time and celebrate them. You don't have to be religious or even the same religion as others to enjoy a meal with friends and family to celebrate the harvest season.

Find out when the farmers' market will have fresh strawberries at the beginning of the season and declare it your own personal Strawberry Day. Celebrate it by having cocktails with friends or simply eating some strawberries on your front porch. The idea is to approach the seasons, and their markers, with a sense of mindfulness and purpose.

Check the almanac

Need a schedule to help mark the passage of time? Get a farmer's almanac or the *Stella Natura*. Both provide readers with the time to do just about everything. Want to know when to sign a contract? Check the almanac? When is the best time to build a fence? Check the almanac. Despite the almanac's insistence on using astrology to guide human behavior, which could lead to a rejection of its advice, there's still something to be said for a season calendar that's still able to tell you when to plant potatoes and buy new clothes.

CONCLUSION

Modern homesteading, at its core, is the simple act of living mindfully within nature while respecting the cycles that govern the world and everything in it. While that may sound overly spiritual and new-agey, ultimately, it's just embracing the world as it is.

By experiencing the seasons, taking a little time and space to separate us from the blatant consumerism of the world economy, and by reaching out to our family, friends, and neighbors, we work hard to achieve a more balanced view of the world.

Homesteaders raise chickens in their backyards, bake their own bread, grow their own herbs, use a sewing machine, plant a garden, and can their own food.

Homesteaders are the DIYers, the makers, and the fence builders. They're modern-day pioneers. They enjoy boots and jeans, and they work hard to separate themselves from the rat race. They enjoy good food, good music, and good books.

But they also do other things. Homesteading is a mindset as much as it is a lifestyle. Sometimes homesteading is just a dream, an ideal, or a goal.

Everyone, no matter their skill level, or what they know, can contribute to the movement. Even the simple act of making a home-cooked meal is an act of homesteading. And in this day and age of fast food and frozen dinners, a good meal can be just as revolutionary and groundbreaking as a yurt in Montana.

Everyone can contribute to the modern homesteading movement, and hopefully this book has provided you with some ideas and methods to make that happen.

Acknowledgments

This is my first, and hopefully not my last, book. But, just in case, I felt it incumbent upon me to thank as many people as I could think of who have helped make this possible.

I'd like to thank my lovely and amazing wife, Melissa, who took many of the photos in this book, for her constant work in shepherding me to write this book, and to do better work every day. She's a constant inspiration, and I hope one day to be good enough to deserve her.

Of course, I'd like to thank my kids, Oliver and Hannah. They're amazing and have been an excellent source of help with everything I do. From making newspaper pots to feeding chickens, they can always be counted on to help out with minimal complaint.

If this book is a success, it's in large part to my editor, Brooke Rockwell, who made me look very, very good in this book. I hope I can buy her a meal soon.

I'd like to thank Abigail Gehring of Skyhorse Publishing for reaching out to me in the first place and for putting up with all my questions and concerns. Thanks to everyone at Skyhorse for even publishing this book; it's an honor.

Thank you Jennifer Hall and Katherine Benoit, who, quite frankly, are the best pair of homesteaders I know. You are an inspiration.

I'm thankful to Chris McLaughlin for answering all of the crazy questions I could think of during the writing process and helping to keep me sane—couldn't have done it without you Snowflake.

Thank you Rene Lee for being a sounding board and a cheerleader.

Larry Kent, formerly with the Onslow County Cooperative Extension Office in North Carolina, allowed me to utterly wreck a quarter-acre section of Onslow County property, and I wouldn't know nearly as much about how not to put a crop in the ground without that chance. Thank you for giving me that opportunity.

Lisa Rayburn, also of the Onslow County Cooperative Extension Office, the best dirt nerd I know, has always been a help. She's always available to answer my weird plant and soil science questions without judgment, no matter how far-fetched they are, and for that I will always be eternally grateful.

Also with the Onslow County Extension Office, I'd like to thank Nicole Sanchez and Robin Seitz, who taught me and my wife so very much.

I'd like to thank Kevin and Kate Justus for being so unbelievably awesome about everything, including being my friends. Love you both!

To John Gullion, my former boss and a current friend, thank you for supporting me and helping me to become a better writer. While I'm at it, I'd also like to thank Tracie Troha. I'll be indebted to both of you for the rest of my life.

Big thanks to Marsha Miller-Howe, Jurlonna Walker, and Linda Borghi for serving as an inspiration and welcoming me to my new home.

For the work they've done with me on the Slow Food Fayetteville Board, I'd like to thank Matt McMahon and Samantha Peterson. And, while I'm at it, I'd like to thank the rest of the board: Carl Pringle, Tiffany Ketchum and Carrie Blackburn McMahon.

Thank you Julian Helms for keeping me sane.

I'd also like to thank Hanah Ehrenreich of Sustainable Sandhills for all the help she's given me and my family since we moved to our new home in Fayetteville, North Carolina.

To my friend Stacie Carames, thank you for supporting me.

Thank you Rabbi Eve Eichenholtz for being such a bright spot in my life and the lives of my family.

I'd like to thank my best friend Kelly Brown for being my best friend.

Thank you Kelsy Howe-Timas for opening your studio and work space at Guiding Wellness so we could host classes.

I'd like to thank my mother- and father-in-law, Jim and Evelyn Nelson, for all of their help through the years.

For being my friends and just all-around lovely human beings, thank you Tiffany Toler, Denise Paul, Adrienne Trego and her husband Daniel DiMaria, Tonya Tallulah Morris, Robert Walker and his wife Nacole, Jacqueline Paige Staab, Pete Swigart, and many, many more.

To Jacqui Sayers, thank you for introducing me to the world of publishing.

I'd like to thank my maternal and paternal grandparents, who served as an inspiration to me.

And, of course, I'd like to thank my mother and father, Randall and Donna Jones, who raised me and my brother and sisters: Josh, Stephanie, and Amanda.

For all of these people named, and for those I've inadvertently left out, thank you for your help, your support, and your inspiration over the years. I would have twice as many vices and half as many virtues without you all. Thank you and I love you.

If I left anyone out, I'm sorry. I promise I'll get you in the next one.

2017 Crop Plan

Crop	Rows	Row Feet Total	Approx. Yield	Sale Price (est.)	Possible Sales
Cantaloupes	3	420	504 melons	$2.50/melon	$1,260
Carrots	2	280	560 carrots (5 per bunch)	$1.50/bunch	$168
Cilantro	3	420	420 bunches	$1/bunch	$420
Cucumbers	4	560	952 pounds	$1/pound	$952
Kale	6	840	1,680 ½-gallon bags	$2/bag	$3,360
Lettuce	3	420	420 ½-gallon bags	$2/bag	$840
Okra	6	840	1,470 pounds	$1.75/pound	$2,573
Potatoes	4	280	560 pounds	$1/pound	$560
Rattlesnake Beans	6	840	252 pounds	$2.50/pound	$630
Roma Tomatoes	6	840	2,100 pounds	$2/pound	$4,200
Spinach	4	560	560 ½-gallon bags	$2/bag	$1,120
Summer Squash/Zucchini	6	840	672 pounds	$1.75/pound	$1,176
Sweet Potatoes	4	280	560 pounds	$1/pound	$560
TOTAL Potential Sales					**$17,819**

APPENDIX 1

Sample Farm Plan

Goal: To secure $6,000 in overall profit through sales of produce

Methods: Grow fourteen crops for sale via a small CSA, direct sales, and sales at the Onslow County Farmers' Market

Lessons learned from the previous year:

1. Increase production
2. Leaffooted plant bugs are the devil
3. Growing trellising is more fun and cheaper than buying it
4. Appearances matter
5. Not having a weed control plan is the same as having a weed growing plan
6. Marketing is easier than you think
7. Keep records
8. Harvests must be sold or processed immediately

Plans of Action/Best Practices

Increasing production—We will grow fourteen crops this year at triple the amounts of the previous year. The crops, despite increasing production, will take up approximately the same amount of space. Other crops will also be grown, but mostly as experiments and market tests. The fourteen crops need to provide about $9,000 in sales to provide necessary profits [$6,000 goal] for the year.

Leaffooted plant bugs and other pest menaces—A handful of chemicals will be used to fight insect infestation (spinosad for potato beetles, Bt for hornworms, etc). In addition, neem oil will be used to combat fungal infections. Regarding the leaffooted plant bugs, however, as they were the biggest problem we faced last year, we will take a two-pronged approach: 1. Use of a kaolin clay mixture on susceptible fruits to discourage feeding by the insects (particularly tomatoes). 2. We will plant about one-eighth acre of black oil sunflower seeds. The sunflowers will serve as a trap plant for the leaffooted plant bugs and also offer a chance to increase income through cut flower sales, if possible. The sunflowers are not considered one of the fourteen crops grown for sale. They are serving primarily as part of an integrated pest management program and as trellising for tomatoes and other crops.

Growing trellising—Getting the materials to make trellises—stakes, strings, cables, etc.—was a problem last season, leading to lost harvests. Use of corn plants, as in the Three Sisters method, provided great trellising for beans. Research shows corn has also been used to trellis other plants, such as cucumbers. This year we'll trellis cucumbers and beans with corn plants. The corn plants will primarily be grown as trellis, and as such aren't considered to be part of the fourteen crops designed to bring in profit. If, however, any of the grains can be sold they will be.

Most of the sunflowers will be grown in a field as a trap for insects. Some varieties, particularly giant types, will be grown with tomatoes to act as trellising for those plants.

I also intend on experimenting with using old okra stalks as trellising, but I'm not expecting much success. However, if it does work, it will save on labor.

Appearances matter—During the 2014 season I learned that appearances matter more than probably anything else. The fruits produced must be pleasing in appearance to sell properly, the field must be pleasing in appearance to market the entire business properly, and to make it easier to work.

Psychologically speaking, a pleasant area to work in boosts morale much more than a mass of weeds. An additional benefit is insect and weed control. If the lot is neat and clean, it reduces insect and weed populations. This will be a big part of prep work this year.

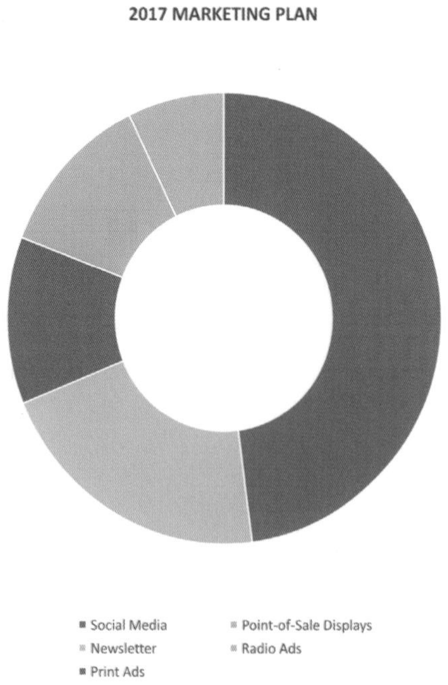

Weed control—Weed control will be accomplished primarily through the use of ryegrass and other cover crops. In addition, mowing and tilling will occur regularly to prevent the rise of weeds between and around beds. Most beds will be covered in plastic. In addition, regular hoeing and hand weeding will augment the process.

Marketing—For the 2014 season, marketing was barely existent. However, we were able to accomplish enough in sales by basic word of mouth to cover costs and break even. For 2015, we're doing a multimedia marketing plan, including social media, an active Internet presence and point-of-purchase displays at the farmers' market. We are also considering producing a newsletter with updates on harvests for subscribers.

Keep records—Record keeping was solid, but unorganized last year. Record keeping this year will consist of spreadsheets for production, sales, and hours worked. They'll be broken down by crop, method of sales, and basic notes for tasks accomplished during hours worked. We'll also continue the extensive journaling process from last year.

Dealing with harvests—By using an extensive planting calendar based on a farmer's almanac, including projected harvest dates, we'll be able to provide CSA subscribers and direct sales customers with the freshest produce with at least a week's notice. Harvests must be collected immediately upon ripening to the correct stage. Any harvest that cannot be sold must be stored immediately. If this isn't possible, then the harvest must be processed (canned or frozen) for home consumption. If this isn't possible, the produce will be donated to needy families and the local homeless shelter. Receipts will be collected for any charity donations with a charge of $1 per pound as a wholesale price for tax purposes.

Plan for next year starting now—We will spend time from November until January clearing weeds from the property, repairing torn plastic, planting a cover crop in between and around beds (annual ryegrass), and buying tools. Planting is scheduled to start at the end of January.

Work schedule—Eighteen hours of labor on the field with an additional six hours of labor for marketing and sales per week, including preparing displays for the farmers' market, maintaining social media accounts, and website production.

Financial breakdown (simple)

Cost estimates

Tiller (used)	$200
Seed	$300
Tools (miscellaneous)	$150
Mower	$80
Ryegrass	$100 (100 pounds)
Sunflower seed	$75
Labor	$1,728

(18 hours a week, $8 an hour, 12-week season)

Chemicals

Bt	$30 (24 ounces)
Neem oil	$25 (16 ounces)
spinosad	$21 (16 ounces)
kaolin clay	$25 (15 pounds)
Miscellaneous	$100
Subtotal:	$2,834
10% cushion	$283
Total projected costs	**$3,117**

Sales goals

CSA (5 baskets, 12-week season)	$1,200
Farmers' market	$2,000
Direct sales	$6,020
Total	**$9,220**
Minus costs	–$3,117
Projected profit	**$6,103**

Crops planned and projected yields

(Figures used for estimates from Louisiana State University AgCenter)

Crop: Cilantro
Rows: 3
Row feet total: 420
Approximate yield: 420 bunches
Sale price (est.): $1 per bunch
Possible sales: $420

Crop: Lettuce
Rows: 3
Row feet total: 420
Approximate yield: 420 bags
Sale price (est.): $2 per bag (½-gallon size)
Possible sales: $840

Crop: Kale
Rows: 6
Row feet total: 840
Approximate yield: 1,680 bags
Sale price (est.): $2 per bag (½-gallon size)
Possible sales: $3,360

Crop: Carrots
Rows: 2
Row feet total: 280
Approximate yield: 560 carrots (sold in 5 carrot bunches, 112 bunches total)
Sale price (est.): $1.50 per bunch
Possible sales: $168

Crop: Potatoes
Rows: 4

Row feet total: 280
Approximate yield: 560 pounds
Sale price (est.): $1 per pound
Possible sales: $560

Crop: Spinach
Rows: 4
Row feet total: 560
Approximate yield: 560 bags
Sale price (est.): $2 per bag (½-gallon size)
Possible sales: $1,120

Crop: Rattlesnake beans (pole)
Rows: 6
Row feet total: 840
Approximate yield: 252 pounds
Sale price (est.): $2.50 per pound
Possible sales: $630

Crop: Summer squash/zucchini
Rows: 6
Row feet total: 840
Approximate yield: 672 pounds
Sale price (est.): $1.75 per pound
Possible sales: $1,176

Crop: Sweet potatoes
Rows: 4
Row feet total: 280
Approximate yield: 560 pounds
Sale price (est.): $1 per pound
Possible sales: $560

Crop: Okra
Rows: 6
Row feet total: 840
Approximate yield: 1,470 pounds
Sale price (est.): $1.75 per pound
Possible sales: $2,573

Crop: Cucumbers
Rows: 4
Row feet total: 560
Approximate yield: 952 pounds
Sale price (est.): $1 per pound
Possible sales: $952

Crop: Cantaloupes
Rows: 3
Row feet total: 420
Approximate yield: 504 melons
Sale price (est.): $2.50 per melon
Possible sales: $1,260

Crop: Roma Tomatoes
Rows: 6
Row feet total: 840
Approximate yield: 2,100 pounds
Sale price (est.): $2 per pound
Possible sales: $4,200

APPENDIX 2

The Basics of Composting

Everyone, whether involved in homesteading or not, should be compositing. Composting takes organic material and allows it to break down naturally, providing a resource for gardens and farms, while preventing extraneous material from taking up space in landfills.

Oddly enough, much of the material that can be composted and broken down into organic material and nutrition for plants and soil, cannot be broken down in landfills. Since material in landfills tends to be sealed away from the oxygen and bacteria that lead to decomposition, the material sits inside the landfill for an indefinite period of time. So while it may seem counterintuitive, taking the material that breaks down easiest out of the landfill is better for the environment in general.

Composting can also save you money. Instead of buying potting soil for your herbs, potted plants, or raised garden beds, you can use compost.

Composting can be done in a homemade compost bin, a professionally purchased bin, or even in a pile in your backyard. Homemade bins can be made of anything. I like to make mine out of untreated pallets. Simply screw four of them together and start filling it. It's even simpler with the professional models and the piles.

As you fill up your compost bin, you'll start stirring in and turning the material. As you do so, you'll add oxygen to the mixture, allowing it to break down faster and more efficiently.

While you may be tempted to put everything in your compost bin, there's actually a surprising number of items you can't. Here's a partial list of things to put in your compost and things to leave out of it.

Add to your compost
Vegetable scraps
Coffee grounds
Bread
Weeds from the garden
Fallen leaves
Grass clippings

Things to avoid
Dog feces
Meat products

Note: If you have chickens, you should put a compost bin inside the chicken run. The compost will draw insects, which can feed the chickens. In addition, the chickens will keep the compost turned as they scratch through. If you put a compost bin in the chicken run, you'll also be able to compost meat, as the chickens will eat it.

APPENDIX 3

Tools

Once you start homesteading, you will suddenly look around and find dozens of projects that need doing yesterday. If you're anything like me, you will suddenly find yourself faced with a shortage of the proper tools.

After a few months, I managed to put together a collection that ensures that no matter the project, I will have all the tools at hand to make short work of it.

Here is my list of tools that every homesteader needs (not including hoes, shovels, and rakes—those should be obvious).

Good hammer—A good claw hammer is the Swiss Army knife of tools. Not only will you use it for its intended purpose, hammering nails, but you will also find yourself using it for the most surprising tasks in a pinch. I have actually used mine for cutting sheet metal, when the quality of the cut edge is not a factor. Hammers are intensely personal tools, but you cannot go wrong with an Estwing, known for its quality. I prefer the lighter twelve-ounce version; it helps prevent fatigue compared to a heavier sixteen-ounce hammer.

Electric circular saw—If you're building anything with wood, you will need to make a cut at some point. You can use a handsaw if you like, but if you have any amount of work to do, you should probably buy a circular saw. I've used a Skil circular saw for about a year now. It's a no-frills tool that never fails to do the job.

Four-foot level—Once your wood is cut, make sure it is put in place level and true. A four-foot-long level will work for just about any project you have, and the extra length will come in handy. It's better to have a level that's longer than you need than one that's too short. There's one from Stanley that's a good option and worth the money, but just about any level with a spirit bubble will do the trick.

Thirty-foot tape measure—This one is pretty self-explanatory. Make sure to get a tape measure that's at least thirty feet long. It's just short enough to be manageable, while long enough to take care of any project short of building a barn from the ground up.

Speed square—Primarily useful for making square cuts, this little gem can be used in lieu of a framing square if necessary. While not as accurate, it is more than adequate for "homestead code." I have discovered that chickens don't seem to mind if the coop is not perfectly square.

Electric drill—While they may not be as convenient as their cordless brethren, an electric drill that may require the use of an extension cord will never leave you with a screw half-way screwed in some really dense wood while you wait for a battery to recharge. Black and Decker makes a good drill, which may not technically be as sturdy as other brands, but you will probably never notice the difference.

Sledge hammer—You may not be planning to do any major demolition soon, but a sledge hammer has dozens of uses, from driving stakes into the ground to driving wedges into logs. I find a two-pound weight to be just the right size: heavy enough to handle most jobs, but light enough for just about anyone to handle. Estwing makes a great one.

Staple gun—Sometimes pounding nails into light wood to hold even lighter wire is not worth the effort. A staple gun means you will never tear strips of wood apart to hold chicken wire in place again. Electric models are available, which can save on hand fatigue, but a manual staple gun will last a long time and you'll never need an extension cord when using it.

Wrenches—Whether you're changing the oil in your car or rebuilding attachments for your tractor, you'll need two good set of wrenches: standard and metric. A set of stubby wrenches by Craftsman will last longer than you will . . . and will get into some tight spaces in the process. Adding a piece of pipe when you need the leverage will make up for the short lengths.

Screwdrivers—Again, you will need screwdrivers for any mechanical issues around the old homestead. Craftsman does it best, and having a set, instead of tools with interchangeable heads, will be worth the extra space it takes up.

There's no such thing as a complete list of tools required for any job, but this is definitely a good start. If you have these items, there's probably no job on the homestead you can't tackle.

APPENDIX 4

Soil Fertility Basics

While there may be a lot of scientific terms bandied about, the basics of soil fertility are not hard to grasp. Soil fertility is determined primarily by the content of macronutrients in the soil that are available for plants to use. It's also determined in part by the content of the micronutrients plants need.

First the macronutrients

The bulk of what plants need, other than sunlight and water, are macronutrients: nitrogen, phosphorous, and potassium. These are the primary nutrients. The intermediate nutrients are sulfur, magnesium, and calcium. Combined with the primary nutrients, the elements compose the macronutrients needed by a plant. The three primary elements are known as N, P, and K, as they are identified by their symbols on the Periodic Table of Elements (N = Nitrogen, P = Phosphorous, and K = Potassium).

Nitrogen is the most volatile of the three elements, and as such is rarely tested for. The thinking most soil labs have is that nitrogen leaves the soil so quickly that by the time the sample gets to the lab, the nitrogen reading won't be accurate. Phosphorous and potassium are not as volatile and tend to remain in the soil longer.

Now the micronutrients

The micronutrients are: iron, boron, manganese, zinc, molybdenum, and copper. To determine how much of these nutrients you need to add to your soil, conduct a soil test. Most soil testing labs will not only pinpoint deficiencies in your soil, but they will also prescribe a methodology to correct them.

APPENDIX 5

Solarizing Raised Beds

Weed control, especially on a small homestead, is going to be a big problem. Most large farm operations rely on a combination of mechanical weed control (aka pulling a cultivator attachment behind your tractor) or chemical weed control, which entails spraying weed killer between the rows. However, homesteading plots are often so small that it's impossible to pull a tractor, and chemical controls tend to go against homesteaders' beliefs.

What do you do when the property refuses to cooperate? Solarize it!

Solarizing a garden bed entails taking a piece of clear plastic, the heavier the better—I prefer at least 3.5 mil plastic (the thickness of the plastic is measured by mil)—and covering the ground with it in order to combat weeds and pests.

The radiation from the sun, and the heat and light, are trapped under the plastic, which cooks weeds, weed seeds, pathogens, insects, and their little eggs, too (insert witchy cackle here).

It has to be done when the weather is right, the hotter the better, and you have to be careful to seal the edges of the plastic.

You can dig a trench around the bed, tuck the plastic into the trench, and then cover it up. I prefer to shovel handfuls of earth onto the edges of the plastic, and then cover the edges with mulch. This gives it a bit more coverage around the edges, looks better, and prevents weeds from coming up on the edges of the bed. If the edges aren't sealed, the moisture and the heat will escape and can actually encourage weed growth.

After about three to four weeks (or more, the University of Florida suggests six weeks) of good, hot weather, the soil is cooked up to six inches into the ground.

Don't worry about earthworms. Research shows they'll actually burrow deeper into the soil to escape the heat, returning when things cool off.

You may be concerned, however, with the fate of beneficial microbes in the soil. Research from the Rodale Institute states that the process will kill beneficial microbes. They suggest using compost, compost tea, manure, and other amendments to the soil as soon as you're finished solarizing it.

Others argue over what type of plastic to use when solarizing beds. Some people use black plastic, while others use a thinner, clear plastic and report higher temperatures. Rodale states that farmers/gardeners get higher temperatures with clear plastic, while the University of Florida states that the thickness of the plastic is only important regarding its ability to withstand your climate.

The only other drawback is the plastic itself. It's made of relatively icky stuff, so you'll have to determine your own comfort level regarding it. Just be sure to recycle it when you're done.

ABOUT THE AUTHOR

Steven Jones is a writer and farmer in North Carolina with two decades of experience as a journalist for television, newspapers, and magazines. Five years ago, he decided to become a homesteader. Despite being the grandson of sharecroppers in Alabama, he found he had none of the skills and know-how to make it happen. So, he attended dozens of classes, conferences, and workshops, and read tons of books, articles, and websites to find out how to do it. He and his wife used this information and their skills to create *From Scratch* magazine, an online publication devoted to homesteading and intentional living. He now operates a four-acre micro farm.

INDEX

Notes